Safety of Repair, Maintenance, Minor Alteration, and Addition (RMAA) Works

T0174012

Safety of RMAA works is an almost uncharted topic of rising importance internationally. Small construction contractors are particularly dependent on RMAA work, especially during times of recession, and they undertake more risks on these jobs than large companies do. This book is based on unique international research and consultancy projects which detail, investigate, and suggest solutions to the specific challenges of safety in RMAA works, based on case studies.

Starting with an overview of safety in the wider construction industries of developed countries, the first half of this book also provides a comprehensive summary of relevant rules, regulations, and the resulting safety performances. The systems in the UK, US, and Hong Kong are described and contrasted, giving the reader an understanding of how different regulatory approaches have yielded a variety of results. From this solid introduction, specific problems observed in RMAA work are examined through case studies, with reference to the underlying cultural and demographic factors, and a variety of practical engineering and management solutions are explored.

This important and practical international work is essential reading for postgraduate students of health and safety in construction, construction project management, or construction in developing countries, as well as policymakers and construction project managers.

Albert P. C. Chan is Chair Professor of Construction Engineering and Management and Head of the Department of Building and Real Estate at Hong Kong Polytechnic University. A chartered construction manager, engineer, project manager, and surveyor by profession, Prof. Chan has worked in a number of tertiary institutions both in Hong Kong and overseas. Prof. Chan's research and teaching interests include project management and project success, construction procurement and relational contracting, construction management and economics, construction health and safety, and construction industry development. He has won a number of prestigious research paper and innovation awards since 1995. Prof. Chan holds an MSc in construction management and economics from the University of Aston in Birmingham, UK, and a PhD in project management from the University of

South Australia. Prof. Chan maintains good links with overseas institutions. He has been an adjunct professor in a number of universities. Prof. Chan was also a founding director of the Construction Industry Institute, Hong Kong, which was a joint research institution developed by industry and academia.

Carol K. H. Hon is currently a lecturer in the construction and project management discipline at the School of Civil Engineering and Built Environment, Science and Engineering Faculty, Queensland University of Technology, Australia. She worked as a postdoctoral fellow in the Department of Building and Real Estate, Hong Kong Polytechnic University. She was a visiting scholar in the Rinker School of Building Construction, University of Florida, USA, in 2010. Her doctoral study has won several prizes and scholarships, including Champion – Construction Industry Council (CIC) Award (Doctoral Group) of the 'Student Project Competition (SPC) 2011' organized by the Hong Kong Institution of Engineers (HKIE) – Safety Specialist Committee (SSC), OSH Best Project Award for the academic year 2011 and OSH Research Scholarship for the academic year 2009, both organized by the Occupational Safety and Health Council of Hong Kong. She has publications in top-tier academic journals and conference proceedings in the field of safety/construction.

Safety of Repair, Maintenance, Minor Alteration, and Addition (RMAA) Works

A new focus of construction safety

Albert P. C. Chan and Carol K. H. Hon

Routledge
Taylor & Francis Group

LONDON AND NEW YORK

First published 2016
by Routledge
2 Park Square, Milton Park, Abingdon, Oxon OX14 4RN

and by Routledge
605 Third Avenue, New York, NY 10017

First issued in paperback 2020

Routledge is an imprint of the Taylor & Francis Group, an informa business.

British Library Cataloguing-in-Publication Data
A catalogue record for this book is available from the British Library

Library of Congress Cataloging in Publication Data
Chan, Albert P. C., author.
Safety of repair, maintenance, minor alteration, and addition (RMAA) works : a new focus of construction safety / Albert P.C. Chan and Carol K.H. Hon.
pages cm
Includes bibliographical references and index.
1. Buildings—Repair and reconstruction—Safety measures. I. Hon, Carol K. H., author. II. Title.
TH3411.C38 2016
690'.22—dc23
2015027028

ISBN 13: 978-0-367-73726-9 (pbk)
ISBN 13: 978-0-415-84424-6 (hbk)

Typeset in Sabon
by diacriTech, Chennai

Contents

Figures

Tables

Foreword

Safety of repair, maintenance, minor alteration, and addition (RMAA) works is of a rising concern to the construction industry in Hong Kong. There is a surge of RMAA works in many developed cities to upkeep aged buildings and improve building energy efficiency and Hong Kong is no exception. RMAA works are very often performed by small- and medium-sized contracting companies which have limited safety training and safety resources. Statistics reveal that the safety of RMAA works has emerged to be a grave concern. In spite of its escalating importance, the safety of RMAA works is a rather uncharted topic and has long been overlooked. Management of RMAA safety deserves greater attention from the academia and practitioners. Such management may result in enhancing safety performance of the construction industry as a whole.

This book delves into topics not commonly addressed by textbooks on construction safety. It records a series of premier research studies related to the RMAA safety together with a critical review of current safety practices in the RMAA sector. The book begins with a general introduction of the RMAA sector, examination of its safety performance, and a review of different accident causation models and safety management related to the RMAA sector. Then, it moves on to discuss specific topics of RMAA safety, including fatal accidents, safety problems and practices, engineering solutions, safety climate, demographic variables and improvement strategies. This book has apparent distinguishing features in that it distils key findings of credible research studies on RMAA safety and presents them in a coherent way to enable readers to grasp a comprehensive picture of the topic.

It is my pleasure to highly recommend this book to academics and practitioners, as well as to students. It is of particular relevance to those who are interested in RMAA safety. You will find this book informative, enlightening and easy to read.

Sr Vincent Ho
President 2014–2015
The Hong Kong Institute of Surveyors

.

Preface

The safety performance of construction in many developed countries has improved remarkably. However, the level of improvement has diminished progressively and some countries have reached a plateau. The challenge for the construction industry today is how to achieve further safety improvement. We perceive that safety of repair, maintenance, minor alteration, and addition (RMAA) works is a new focus of construction safety. Addressing the safety problems of RMAA works will further improve construction safety performance to a new level.

Safety of RMAA works is a niche area. Although there are many textbooks on construction safety, those which are written specifically to address safety of RMAA works are scarce. This may be due to the fact that many RMAA works are done by small- and medium-sized contractors and they are not as richly resourced as the large contractors. Accident statistics for the RMAA sector are not commonly available in the public domain. We are fortunate to have the support of the Labour Department of the Hong Kong government to provide the accident statistics of RMAA works and analyses of fatal accidents of RMAA works in Hong Kong.

The Construction Health and Safety Research Group at the Hong Kong Polytechnic University has completed a number of research projects in this area. It is high time to turn years of research efforts into fruition. Despite the fact that this book is a documentation of a long-term research endeavour, it is presented in a way to fit a wide audience. This book provides the novice reader with an introduction of RMAA works and their related safety issues. It also provides in-depth research findings for graduate students and researchers.

This book consists of nine chapters. Chapter 1 provides an overview of the rising importance of the RMAA sector in a number of countries. Chapter 2 summarises the safety performance of the construction industry in developed countries. Chapter 3 is about accident causation models and safety management. Chapter 4 provides a detailed fatal case analysis of RMAA works. Chapter 5 provides an account of safety problems and practices of RMAA works. Chapter 6 introduces some engineering solutions to RMAA works. Chapter 7 shows the measurement of safety climate of

RMAA works. Chapter 8 is about the relationships between demographic variables and safety of RMAA works. Chapter 9 provides strategies for improving safety of RMAA works.

The financial support from the Research Grants Council of the Hong Kong Special Administrative Region, China (RGC Project No. PolyU5103/07E) is gratefully acknowledged. The authors are also indebted to the Hong Kong Institute of Surveyors for providing a sponsorship for writing this book. It is hoped that this book will benefit the construction industry and contribute to saving many lives.

1 Composition of the construction market in developed societies

Introduction

The construction industry in a broad sense includes activities of new construction works, and repair, maintenance, minor alteration and addition (RMAA) works. New construction works include building, civil engineering and infrastructure works. RMAA works are construction activities conducted to existing structures and the built environment. Construction markets in developed countries have been changing. In some places, new construction market shrinks but RMAA market expands. This chapter will examine the composition of the construction industry in developed societies, explaining why the RMAA sector of the construction industry deserves attention of society.

Definitions of repair, maintenance, minor alteration and addition (RMAA) works

Repair, maintenance, minor alteration and addition (RMAA) works is part of the whole building life cycle. The building life cycle begins when the client decides to develop and terminates when the asset is disposed (BS3811:1993). Repair and maintenance is the longest process in the whole building life cycle as compared with other processes such as acquisition and demolition. Once a building is constructed, it begins to deteriorate. Repair and maintenance is required to ensure that the fabric and facilities of the building function at an acceptable level (Chan et al., 2010).

Referring to the British Standard BS3811:1993, 'Glossary of Terms in Terotechnology', maintenance is defined as 'the combination of all technical and administrative actions, including supervision actions, intended to retain an item in, or restore it to, a state in which it can perform a required function'. This definition involves two processes: retaining and restoring. Retaining is work carried out in anticipation of failure whereas restoring is work carried out after failure. The former is 'preventive maintenance' while the latter is 'corrective maintenance' (Wordsworth, 2001). Repair is defined in the same standards as 'that part of corrective maintenance in which manual actions are performed on the item'. Repair and maintenance include all the activities carried out to sustain the performance of both the building fabric and the building services installations (Chan et al., 2010).

Repair and maintenance should also include a reasonable element of improvement (Wordsworth, 2001). For example, worn-out building services are to be replaced with the latest systems. However, to retain or to restore under the definition of maintenance suggests the initial standard be taken as the proper basis. If the required function is altered or raised in standard where retaining or replacing the existing components may not be able to meet the new standard, minor alteration and addition works would be required (Chan et al., 2010).

RMAA works are different from major redevelopment works. RMAA works are undertaken to maintain the existing functional life of a building whereas redevelopment works create a building with new functionalities. Major redevelopments may involve large-scale improvement or conversion which alters or increases the utility of the building (Catt & Catt, 1981).

The Labour Department of the Hong Kong government defines RMAA as those minor works such as construction projects for village-type houses in the New Territories, minor alterations, repairs, maintenance and interior decoration of existing buildings (Labour Department – Hong Kong, 2008). The Report on the Quarterly Survey of Construction Output of the Census and Statistics Department of the Hong Kong government (Census and Statistics Department – Hong Kong, 2009) defines RMAA works as the 'construction works at locations other than sites' which comprises general trades and special trades. General trades include 'decoration, repair and maintenance and construction works at minor work locations such as site investigation, demolition, and structural alternation and addition works'. Special trades include 'carpentry, electrical equipment, ventilation, gas and water fitting installation and maintenance etc.' (Census and Statistics Department – Hong Kong, 2009).

RMAA works in a broad sense include any improvements to existing buildings to maintain their viabilities or enhance their energy efficiencies. Refurbishment, retrofitting, renovation and remodelling commonly involve special trade construction activities to improve existing buildings for better energy efficiency or heritage protection. They are sometimes used interchangeably in different contexts. These adaptation activities in the built environment have become more and more important due to climate change. Examples include installation of solar panels, solar hot water systems, and wall and ceiling insulation. They may be considered periphery activities to the construction industry and somewhat different from traditional RMAA works but we believe that it would be beneficial to include these activities in this book to reflect the recent development of the RMAA sector and thus provide more insightful and up-to-date discussion.

Changing composition of the construction market

Construction markets of developed societies have a different composition structure from those of developing ones. Unless there are massive new development plans, the proportion of new construction works in the construction markets of developed societies would be rather static whereas that of RMAA works would sustain. During economic downturn, the importance of RMAA works may even be greater than new construction works. Investments for new construction projects would likely be stopped but the need for repair and maintenance works would be quite stable or even increase. In fact, rolling out RMAA projects has been used as a strategy by some governments to create immediate job opportunities to boost the economy. For example, the Development Bureau of the Hong Kong government rolled out HKD 8.56 billion (approximately USD 1.1 billion) of RMAA works in 2009/2010 after the financial tsunami to stimulate the economy (Development Bureau – Hong Kong, 2008). Similarly, the Australian government announced an AUD 2.8 billion (approximately USD 2 billion) home insulation program in 2009 to create jobs for home retrofitting so as to lessen the hit to the Australian economy (Hanger, 2014).

Rising awareness of sustainability also boosts the volume of RMAA works to a new height. Some refurbishment and retrofitting works are carried out so as to switch to renewable energy or have better energy efficiency. Refurbishment and retrofitting of existing buildings has huge potential for reducing greenhouse gas emission. According to Climate Works Australia (2010), retrofitting of existing buildings is one of the recommended ways to improve energy efficiency and reduce greenhouse gas emission. With reference to the UK Green Building Council (2013, 2014), retrofitting existing domestic homes is vital to achieve national carbon emission targets. According to the Chartered Institute of Building (2011), the number of domestic and nondomestic buildings in the United Kingdom is almost 30 million. In order to meet the carbon targets set by the UK government, it is anticipated that around 28 million buildings (including 25 million homes) are required to be retrofitted by the end of 2050. Up to 85% of housing that will exist in 2050 has already been built. Energy consumption of domestic buildings alone accounts for 27% of all UK carbon emissions. There are approximately 1.8 million nondomestic buildings in the United Kingdom. These are currently responsible for roughly 18% of the country's total carbon emissions. Improving the energy efficiency of these buildings will substantially reduce the amount of total carbon emissions.

Homeowners as well as developers are motivated to carry out refurbishment and retrofitting works to save rising costs of their electricity bills and protect the environment. Improving building energy efficiency will reduce the money spent on electricity. Some governments provide financial incentives to encourage refurbishment and retrofitting works. To achieve their renewable energy target and reduce greenhouse gas emission, the Australian government has set up schemes to encourage green refurbishment and retrofitting. For instance, the Queensland government provides feed-in-tariff financial incentives for homeowners to install solar panels on their rooftops (Department of Energy and Water – Queensland Government, 2014). Refurbishment and retrofitting programs are in place in Melbourne to improve energy efficiency of commercial buildings. The 1,200 Buildings program in Melbourne aims to help building owners, managers and facility managers to improve energy and water efficiency and to reduce waste to landfills of commercial buildings in Melbourne (Melbourne Council, 2015). With these financial incentives, it is expected that retrofits of private office buildings will bring about 10,000 jobs annually to the construction industry of Australia (Climate Works Australia, 2010).

Aging building is another reason why RMAA works become increasingly important to the construction market in developed societies. Proper repair and maintenance will help preserve building value and prolong building life. Since dilapidated buildings endanger public safety, some governments mandatorily require building owners to carry out regular building repair and maintenance (Development Bureau – Hong Kong, 2010). Because of this mandatory requirement, volume of RMAA works continues to expand in many developed societies. As in Hong Kong, buildings aged 30 years or more (except domestic buildings not exceeding three storeys) are required to undergo inspection and carry out necessary repair works of the common parts, external walls and projections or signboards of the buildings. Buildings aged 10 years or above (except domestic buildings not exceeding three storeys) are required to undertake inspection and carry out repair works to all windows of the buildings (Development Bureau – Hong Kong, 2010).

Aging building has become a common issue in developed societies. We can expect an increasing importance of RMAA works in the foreseeable future. By the time the Institute for Building Efficiency (2010) was written, at least 50% of the buildings to be found in 2050 had already been built and about 72% of floor space in the United States belonged to buildings over 20 years old. According to the Chartered Institute of Building (2011), the United Kingdom has the oldest domestic building stock among developed societies. About 8.5 million buildings are over 60 years old (pre-1944: 38%, 1945–1984: 46%, 1985 onwards: 16%). About 75% of the nondomestic building stock are more than 25 years old, while nearly one-third are over 70 years old (pre-1940: 31%, 1940–1985: 46%, 1985 onwards: 23%). A steady demand for repair, maintenance and retrofitting works would be needed to preserve these building stocks or make them more sustainable.

Composition of the construction market in the United States

The repair and maintenance sector is expected to increase 9% from 2012 to 2022 in the United States (Bureau of Labor Statistics – US, 2014). A number of reasons lead to expected expansion of the repair and maintenance sector. According to the Bureau of Labor Statistics – US (2014), more home sales will create more renovation and remodelling works. There will be upgrades and renovation demand for old properties. Demographic changes may also increase the demand for repair and maintenance works. As a large baby-boom population comes to retirement age, it is anticipated that many of them will invest in renovations to accommodate their future living needs and allow them to remain in their homes following retirement. In addition, RMAA works continues to expand during economic recession. The RMAA sector in the United States has been expanding after the credit crisis. During the economic recession, fewer new buildings have been built because of capital constraint. Consequently, more existing buildings have been remodelled for sale or retrofitted to green buildings to save mounting energy expenses (US Bureau of Labor Statistics, 2010a, 2010b).

Many RMAA works in the United States are related to improving energy efficiency of buildings. In the United States, Property-Assessed Clean Energy (PACE) provides financial assistance to encourage energy efficiency retrofitting works. PACE is a means of financing energy efficiency upgrades or renewable energy installations for buildings (US Department of Energy, 2015). Examples of upgrades range from adding more attic insulation to installing rooftop solar panels.

PACE was started in 2008. Pilot PACE programs were conducted in California, Colorado, and New York on energy efficiency upgrades to single family homes. The first actual PACE program was carried out in Berkeley, California, and the state passed the first PACE-enabling legislation in 2008. Under the program, the municipal government will collect money from investors to fund the PACE program by issuing a specific bond. It will then lend out money to property owners for energy retrofitting. The loans are to be repaid over an agreed term (typically 15 or 20 years) via an annual assessment on their property tax bills. PACE contributes to an increasing volume of retrofitting works.

Composition of the construction market in the United Kingdom

Repair and maintenance works accounted for 48% of the construction market in the United Kingdom in 2009 (Office for National Statistics – UK, 2010) and it accounted for 38% of the construction market in 2014 (Office for National Statistics – UK, 2015). As shown in Figure 1.1, repair and maintenance is the largest (38%) work category of the construction industry when compared with, respectively, infrastructure (11%), housing (22%) and other new work (29%).

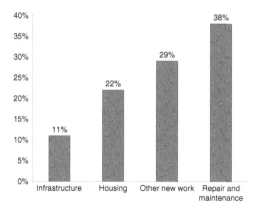

Figure 1.1 Distribution of construction market in the United Kingdom in 2014.
Source: Office for National Statistics – UK (2015).

With the Climate Change Act 2008, the United Kingdom took a world-leading role in setting binding carbon reduction targets, laying out a bold ambition to reduce emissions by 80% by 2050 (UK Green Building Council, 2013). Like other developed countries, the United Kingdom faces the need to reduce greenhouse gas emission to meet its 2050 climate change targets. The domestic sector accounted for 24% of total emissions in the United Kingdom. Housing stock of the United Kingdom is among the most inefficient in Europe and at least 80% of the homes that will be standing in 2050 have already been built (UK Green Building Council, 2013). There are nearly 30 million buildings (domestic and nondomestic) in the United Kingdom. Approximately 28 million of these (including 25 million homes) are required to be retrofitted by the end of 2050 if the carbon targets are to be met (Charted Institute of Building, 2011). Retrofitting existing energy inefficient buildings will bring about some GBP 7 billion (approximately USD 11 billion) of investment annually and create up to 250,000 jobs by 2030 (Charted Institute of Building, 2011). The UK government launched the Green Deal in January 2013. This policy allows participants to carry out energy-efficient improvements to their properties such as insulation at no installation cost but repay through their energy bills. The UK government anticipated the Green Deal would support the retrofit of 14 million homes by 2020 (UK Green Building Council, 2014). Whereas the effectiveness of the policy would take time to reveal, the policy itself would boost the volume of retrofitting and refurbishment works in the United Kingdom.

Composition of the construction market in Hong Kong

The RMAA sector of the construction industry in Hong Kong is often overlooked because a vast majority of RMAA projects are small in size and undertaken by small-sized contractors. Statistics show that the RMAA sector plays an increasingly important role in the construction market of Hong Kong.

As shown in Table 1.1, the RMAA sector increased from 23.5% of the total construction volume in 1998 to 53.5%, the highest ever, in 2006 (Hon, Chan, & Wong, 2010). The increase was more than onefold. From 2007 to the latest figure in 2013, the RMAA sector accounted for an average of 44.2% of the construction market in Hong Kong. With the gradual rollout of 10 infrastructure projects in Hong Kong since 2010, the total construction volume has substantially increased. It is noteworthy that the declining trend of the percentage of RMAA works to the total construction market is mainly due to the expansion of the construction market. The nominal value of the RMAA sector has been expanding at a higher rate than that of the whole construction industry. Value of the RMAA sector increased 13% between 2012 and 2013 whereas the total construction market increased only 9.4% in the same period.

Launching RMAA projects can be an effective short-term strategy to create job opportunities. The economy of Hong Kong was hit by the devastating effect of the global financial crisis in 2008. The Hong Kong government decided to launch more RMAA works to provide immediate employment opportunities in the construction industry (Development Bureau – Hong Kong, 2008). A total of HKD 8.56 billion (approximately USD 1.1 billion) was spent by the government on RMAA works in the financial year of 2009/2010, creating 1,600 jobs in the construction industry (Development Bureau – Hong Kong, 2008). Examples of the RMAA works undertaken are refurbishing external walls of 50 government buildings, renovating aged protective surfaces of 500 slopes, installing and retrofitting energy-efficient facilities for government departments, and putting green roofs on 40 government buildings (*The Standard*, 2009).

It is expected that the RMAA sector will become increasingly important to the construction industry of Hong Kong. Aging buildings are a problem in Hong Kong. In 2006, around one-third of the housing blocks in Hong Kong were built more than 20 years ago (Chan et al., 2006). The Hong Kong government has launched the Mandatory Building Inspection Scheme (MBIS) and Mandatory Window Inspection Scheme (MWIS) in 2010 (Development Bureau – Hong Kong, 2010) to address the problem of aging buildings.

Under the MBIS scheme, each year the Buildings Department of Hong Kong will select 2,000 private buildings aged 30 years or more to undergo building inspection. Domestic buildings not exceeding three stories are exempted. The selected buildings are required to carry out a thorough safety inspection and necessary repair and maintenance work. After that, building inspections are needed every 10 years (Development Bureau – Hong Kong, 2010). The MWIS stipulates that windows of private buildings aged 10 years or more require safety inspection. Windows of those domestic buildings not exceeding three stories are exempted from this requirement. Each year around 5,800 private buildings will be selected for carrying out safety inspections for windows. Necessary window repair and maintenance

Table 1.1 Gross value of construction work at current market prices (1998–2013) (unit: HKD million at current prices; USD 1 = HKD 7.8)

	1998	1999	2000	2001	2002	2003	2004	2005	2006	2007	2008	2009	2010	2011	2012	2013
Residential (A)	48,761	56,225	51,920	41,774	36,503	28,612	20,085	16,945	15,518	16,064	20,613	22,804	22,381	26,083	37,501	38,186
Nonresidential (B)	33,866	20,455	17,407	16,026	16,502	18,243	17,425	17,060	14,161	17,289	17,287	16,938	18,206	21,014	23,453	22,768
Civil Engineering (C)	19,349	16,873	20,583	24,491	21,358	20,710	19,044	14,686	12,311	10,123	10,934	12,516	20,388	30,254	43,067	52,367
Total Construction Investment (A+B+C)	101,975	93,553	89,910	82,290	74,362	67,564	56,553	48,691	41,990	43,476	48,834	52,258	60,974	77,351	104,021	111,766
Repair, Maintenance, Minor Alteration and Addition (D)	31,341	32,884	32,161	31,696	31,638	31,468	36,618	42,160	48,240	49,390	50,765	48,686	49,966	51,184	57,428	64,809
Total Construction Market (A+B+C+D)	133,316	126,437	122,071	113,986	106,000	99,032	93,171	90,851	90,230	92,866	99,599	100,944	110,940	128,535	161,449	176,575
Percentage of RMAA Works to Total Construction Market (%)	23.5	26.0	26.3	27.8	29.8	31.8	39.3	46.4	53.5	53.2	51.0	48.2	45.0	39.8	35.6	36.7

Note: Data sourced from *Report on the Quarterly Survey of Construction Output, Tables 1A and 3, Census and Statistics (CS&D) Department, Hong Kong* (Census and Statistics Department – Hong Kong, 2009).

The CS&D named this figure as 'locations other than sites' which refers to 'works at locations other than construction sites includes minor new construction activities and renovation works at erected buildings and structures; and electrical and mechanical fitting works at locations other than construction sites.'

Source: Census and Statistics Department – Hong Kong (2009).

works will be required. After that, window inspections are needed every five years (Development Bureau – Hong Kong, 2010).

In addition, the Minor Works Control System was implemented on 31 December 2010 (Buildings Department – Hong Kong, 2011). The purpose of the Minor Works Control System is to simplify and speed up the approval procedures of minor RMAA works. To come together with the implementation of the MBIS, MWIS and Minor Works Control System, the Hong Kong government provides subsidy schemes and technical assistance to building owners to support them carrying out repair and maintenance works for their aged buildings (Legislative Council – Hong Kong, 2011). All these initiatives contribute to the continuous expansion of the RMAA sector in Hong Kong.

Chapter summary

This chapter examines the changing composition of the construction industry in developed societies, in particular the United States, the United Kingdom and Hong Kong. It is expected that the number of RMAA works in the construction industry will continue to expand. The RMAA sector has largely been overlooked when the economy is good. However, it has become more and more important for a number of reasons: for example, the government wants to create jobs by launching repair and maintenance projects, more and more aging buildings, rising concern for sustainability and energy efficiency, and implementation of mandatory building inspection schemes. In view of these factors, it is expected that the importance of the RMAA sector in developed societies will continue to increase.

References

Buildings Department – Hong Kong. (2011). *Minor Works Control System: Full Implementation on 31 December 2010*. Retrieved from www.bd.gov.hk/english/services/index_buildingAmendent.html.

Bureau of Labor Statistics – US. (2010a). *Career Guide to Industries, 2010-11 Edition, Construction*. Bureau of Labor Statistics, US Department of Labor. Retrieved from www.bls.gov/oco/cg/cgs003.htm.

Bureau of Labor Statistics – US. (2010b). *Occupational Outlook Handbook, 2010–11 Edition, Maintenance and Repair Workers, General*. Bureau of Labor Statistics, US Department of Labor. Retrieved from www.bls.gov/oco/ocos194.htm.

Bureau of Labor Statistics – US. (2014). *General Maintenance and Repair Workers, Occupational Outlook Handbook*. Bureau of Labor Statistics, US Department of Labor. Retrieved from www.bls.gov/ooh/installation-maintenance-and-repair/general-maintenance-and-repair-workers.htm#tab-6.

Catt, R., & Catt, S. (1981). *The Conversion, Improvement and Extension of Buildings*. Bath, Great Britain: The Pitman Press.

Census and Statistics Department – Hong Kong. (2009). *Report on the Quarterly Survey of Construction Output*. Retrieved from www.censtatd.gov.hk/products_and_services/products/publications/statistical_report/commerce_and_industry/index_cd_B1090002_dt_detail.jsp.

Chan, A. P. C., Wong, F. K. W., Chan, D. W. M., Yam, M. C. H., Kwok, A. W. K., Yiu, E. C. Y., Chan, E. H. W., Lam, E. H. W., & Cheung, E. (2006). A research framework for investigating construction safety against fall of person accidents in residential building repair and maintenance works. In D. Fang, R. M. Choudhry & J. W. Hinze (Eds.), *Proceedings of CIB W99 International Conference on Global Unity for Safety and Health in Construction* (pp. 82–90). Beijing: Tsinghua University Press.

Chan, A. P. C., Wong, F. K. W., Yam, M. C. H., Chan, D. W. M., Hon, C. K. H., Wang, Y., Dingsdag, D., & Biggs, H. (2010). RMAA safety performance – How does it compare with Greenfield projects. In P. Barret, A. Dilanthi, R. Haigh, K. Keraminiyage & C. Pathirage (Eds.), *Proceedings of the CIB World Congress 2010, Building a Better World* (p. 154). Salford Quays, UK: CIB – International Council for Research and Innovation in Building and Construction.

Charted Institute of Building. (2011). *CIOB Carbon Action 2050: Buildings under Refurbishment and Retrofit.* Bracknell, UK: The Chartered Institute of Building.

Climate Works Australia. (2010). *Low Carbon Growth Plan for Australia.* Clayton, Victoria, Australia: Climate Works Australia.

Department of Energy and Water – Queensland Government. (2014). Solar Bonus Scheme [Press release]. Retrieved from www.dews.qld.gov.au/energy-water-home/electricity/solar-bonus-scheme/frequently-asked-questions.

Development Bureau – Hong Kong. (2008). Immediate Measures Proposed to Assist Construction Industry [Press release]. Retrieved from www.devb.gov.hk/en/secretary/press/press20081125a.htm.

Development Bureau – Hong Kong. (2010). *Mandatory Building Inspection Scheme and Mandatory Window Inspection Scheme.* Retrieved from www.bd.gov.hk/english/services/index_MBIS_MWIS.html.

Hanger, I. (2014). *Report of the Royal Commission into the Home Insulation Program.* Australia: Royal Commission into the Home Insulation Program. Retrieved from http://apo.org.au/files/Resource/hirc_reportoftheroyalcommissionintothehomeinsulationprogram_sep_2014.pdf.

Hon, C. K. H., Chan, A. P. C., & Wong, F. K. W. (2010). An empirical study on causes of accidents of repair, maintenance, minor alteration and addition works in Hong Kong. *Safety Science, 48*(7), 894–901.

Institute for Building Efficiency. (2010). Why focus on existing buildings. Retrieved from www.institutebe.com/Existing-Building-Retrofits/Why-Focus-On-Existing-Buildings.aspx.

Labour Department – Hong Kong. (2008). *Accidents in the Construction Industry of Hong Kong (1998–2007).* Labour Department, Government of Hong Kong. Retrieved from www.labour.gov.hk/eng/osh/pdf/AccidentsConstructionIndustry1998-2007.pdf.

Legislative Council – Hong Kong. (2011). *Legislative Council Panel on Manpower – Hong Kong's Occupational Safety Performance in the First Half of 2010 [LC Paper No. CB(2)814/10-11(04)].* Retrieved from www.legco.gov.hk/yr10-11/english/panels/mp/papers/mp0120cb2-814-4-e.pdf.

Melbourne Council. (2015). 1200 Building [Press release]. Retrieved from www.melbourne.vic.gov.au/1200buildings/Pages/Home.aspx.

Office for National Statistics – UK. (2010). *Construction Statistics Annual 2010, UK.* Office for National Statistics, UK. Retrieved from www.statistics.gov.uk/downloads/theme_commerce/CSA-2010/Opening%20page.pdf.

Office for National Statistics – UK. (2015). *Statistical Bulletin: Output in the Construction Industry. December and Q4 2014.* Retrieved from www.ons.gov.uk/ons/dcp171778_395092.pdf.

The Standard. (14 January 2009). Hong Kong, Kai Tak takes off. Retrieved from www.thestandard.com.hk/news_detail.asp?pp_cat=30&art_id=76986&sid= 22235128&con_type=3

UK Green Building Council. (2013). *Retrofit Incentives: Boosting Take-up of Energy Efficiency Measures in Domestic Properties*. UK Green Building Council. Retrieved from www.ukgbc.org/resources/publication/uk-gbc-task-group-report-retrofit-incentives.

UK Green Building Council. (2014). *A Housing Stock Fit for the Future: Making Home Energy Efficiency a National Infrastructure Priority*. UK Green Building Council. Retrieved from www.ukgbc.org/resources/publication/housing-stock-fit-future-making-home-energy-efficiency-national-infrastructure.

US Department of Energy. (2015). *Property-Assessed Clean Energy Programs*. US Department of Energy. Retrieved from http://energy.gov/eere/slsc/property-assessed-clean-energy-programs.

Wordsworth, P. (2001). *Lee's Building Maintenance Management* (4th ed.). Oxford: Blackwell Science.

2 Construction safety performance in developed societies

Introduction

This chapter introduces different kinds of safety performance measurement and reviews the safety performance of the construction industries in the United States, the United Kingdom and Hong Kong. Likewise, the safety regulations governing these jurisdictions have also been reviewed. The safety performances of the RMAA sectors in these places are highlighted in particular. Safety performance of the RMAA sector is often difficult to benchmark. There are no or only limited accident statistics available in the public domain for the RMAA sector.

Occupational safety and health regulatory bodies often measure safety performance with incident rate. This allows across-the-board comparison of different industries and sectors. It is often difficult to obtain an accurate estimate of the number of practitioners in the RMAA sector and thus difficult to calculate the incident rate. The difficulties of benchmarking safety performance of RMAA works will be discussed in this chapter.

Safety performance measurement

To assess safety performance, we need to have measurement indicators. There are two main types of safety performance measurement indicators, namely lagging indicators and leading indicators. Lagging indicators are defined by Australian Constructors Associations (2015) as 'events that have already occurred that cause harm to the people that work in an organization that are measured as an indicator of safety performance' whereas leading indicators are defined by Australian Constructors Associations (2015) as the 'proactive measures that organizations undertake to assist in improving their safety outcomes'.

Traditionally, safety performance is measured by outcome-based lagging safety performance indicators such as fatality rates, lost time injury rates and medical injury rates. In the United States, safety performance is largely measured by metrics like the Occupational Safety and Health Administration

(OSHA) recordable injury rate (RIR); days away, restricted work, or transfer (DART) injury rate; or the experience modification rating (EMR) on workers' compensation (Hinze, Thurman, & Wehle, 2013).

In Australia, the common lagging indicators for safety performance measurement are first aid injury frequency rate (FAIFR), fatality incidence frequency rate (FIFR), lost time injury frequency rate (LTIFR), medically treated injury rate (MTIR), non–medically treated injury rate (NMTIR), notifiable dangerous occurrence rate (NDOR), non-injury incident or near miss/near hit (NII), return to work rate (RTWR), workers' compensation claim rate (WCCR) and workers' compensation premium rate (WCPR) (Biggs, Dingsdag, Kirk, & Cipolla, 2010).

The problem of using lagging indicators is that they are reflecting the past, reactive and have low predictability of future accident occurrence (Australian Constructors Associations, 2015). Actions relating to lagging indicators are often negative because action will be taken only when the lagging indicators show that performance is below expectation. Besides, lagging indicators do not provide any clues why the performance is below expectation.

Unlike lagging indicators, leading indicators measure the performance of the safety processes (Hinze et al., 2013). They are more useful in locating the safety problems and identifying actions to be taken. Leading indicators would help prioritise the efforts to be spent on improving safety. There are two types of leading indicators, passive and active. Passive indicators reflect the safety performance of a company on a macro scale but not short-term and cannot be changed quickly. Examples of passive leading indicators in the United States include (Hinze et al., 2013):

- number or proportion of the managerial staff certified for 10 hours (or 30 hours) OSHA safety training;
- number or proportion of front-line staff certified for 10 hours (or 30 hours) OSHA safety training;
- number or proportion of subcontractors selected by using safety as a selection criterion.

In contrast, active leading indicators can be changed quite easily in a short time. Examples of active leading indicators in the United States include (Hinze et al., 2013):

- number or proportion of site tool-box meetings that site supervisors/project managers joined;
- number or proportion of site pre-task planning meetings that site supervisors/project managers joined;
- number or proportion of safety compliance on safety inspections.

Safety performance measurements of new construction works using lagging indicators are well established. More and more contracting companies have taken a step further to use leading indicators together with lagging indicators to more accurately reflect their safety performance.

Safety performance measurement of RMAA works

Safety performance of the RMAA sector is not available in most of the jurisdictions. The RMAA sector is often overlooked and so is its safety performance. Besides, it is also difficult to measure safety performance of RMAA works. Safety performance of the RMAA sector may not be accurately measured by commonly used lagging indicators because of three main reasons: 1) underreporting of minor injuries, 2) no exact match of activity classification by government statistics department with activity classification by accident reporting, and 3) lack of information on the number of RMAA practitioners.

Underreporting is common for minor injuries in the construction industry (Chan et al., 2010; Moore, Cigularov, Sampson, Rosecrance, & Chen, 2013). This is particularly the case for the RMAA sector. Many RMAA projects are small in scale and thus subject to less stringent monitoring requirements (Chan et al., 2010). For example, construction projects in the United Kingdom are required to notify the Health and Safety Executive before commencement if they will exceed 30 days and involve construction activities exceeding 500 man-days. Small-sized RMAA projects with short duration are exempted (Health and Safety Executive, 2006). Minor injuries of these projects may not have been reported.

In Hong Kong Clause 56 of the Construction Sites (Safety) Regulations stipulates that contractors do not need to report to the Commissioner for Labour upon construction project commencement if the construction activities will be finished within 6 weeks or no more than 10 employees are engaged in the project throughout the whole process. Thus, the Labour Department as a regulatory government body in Hong Kong is unable to check the safety performance of these projects until a fatal accident happens. Injury reporting heavily relies upon the RMAA contractors. According to Cheung (2005), it is common not to report minor RMAA work injuries to the Labour Department of Hong Kong. The Labour Department cannot track down small RMAA projects until these projects have serious industrial accidents. Underreporting of injuries in the RMAA sector has made data collection for safety performance measurement of this sector very difficult.

Construction activities related to the RMAA sector are often not well defined by the statistics department of most jurisdictions. The value of work relating to RMAA is not readily available. Even if it is, the categorizations of work activities are often not in alignment with that of the accident reporting. This makes accurate measurement of RMAA safety performance difficult.

For example, in Hong Kong, the Labour Department (2008) provides a definition of RMAA works and the statistics of the number of accidents occurred in the RMAA sector from 1998 to 2007. However, the Census and Statistics Department does not have exactly the same categorization of construction activities. The closest figure representing RMAA works provided by the Census and Statistics Department is 'Locations other than sites' which is defined as 'Works at locations other than construction sites,' which includes 'minor new construction activities and renovation works at erected buildings and structures; and electrical and mechanical fitting works at locations other than construction sites' (Census and Statistics Department – Hong Kong, 2009). The same problem occurs in the United Kingdom. The Office for National Statistics has different construction activity classifications which are not exactly in line with the accident reporting. Exact match comparison of the volume of work in the RMAA sector and the number of accidents in the RMAA sector cannot be done easily.

Another problem of measuring safety performance of the RMAA sector is lack of employment figures for the RMAA sector in most jurisdictions. Accident rate per a number of workers is one of the most common lagging indicators of safety performance in the construction industry. To compile this, both the numbers of accidents and the number of workers would be required. Safety performance of the RMAA sector is not available in most of the jurisdictions. One of the reasons is lack of information on the number of workers in this sector. For example, in Hong Kong, the accident rate per 1,000 workers is the safety performance indicator for all industries. However, such information is not available for the RMAA sector because there is no official and accurate record of the number of RMAA workers in Hong Kong. RMAA projects are usually undertaken by small-sized contractors with only a few direct labourers. Many RMAA projects involve temporary workers with a high turnover rate. With short project duration and flexible employment, it is very difficult for the government to gather accurate employment figures or to force the contractors to report on such figures (Chan et al., 2010).

Difficulties of measuring safety performance of the RMAA sector have been well recognized by the industry practitioners in Hong Kong (Chan et al., 2010). Some suggested benchmarking safety performance of the RMAA sector with other sectors of the construction industry by the accident rate per value of construction output. The problem of using accident rate per value of construction output as the indicator for safety performance is that different sectors require different degrees of labour intensiveness. RMAA works tend to require a far greater number of workers than for the same amount of work done in new construction projects. With more man-hour inputs, the number of injuries or accidents tends to be higher. Meanwhile, the RMAA sector is lacking a precise measurement of safety performance to enable cross-sector comparisons. For continuous safety improvement of the

RMAA sector, it is indispensable to develop a reliable safety performance indicator which allows comparison across sectors and countries. More effort should be devoted to accurately capture the employment figure and the number of accidents for the RMAA sector (Chan et al., 2010).

Construction safety performance in the United States

According to the US Bureau of Labor Statistics (BLS), the US national average of nonfatal injuries and illness rate was 3.5 cases per 100 equivalent full-time workers in 2013 (Bureau of Labor Statistics – US, 2014a). Manufacturing ranked the first at 4.0 cases per 100 equivalent full-time workers. Natural resources and mining ranked the second at 3.9 cases per 100 equivalent full-time workers. Construction was ranked the third at 3.8 cases per 100 equivalent full-time workers. Out of 3,929 worker fatalities in private industry in 2013, 796 or 20.3% were in construction – that is, one in five workers who died in 2013 were in construction. The leading causes of worker deaths on construction sites were falls, followed by struck by object, electrocution, and caught in between. These 'Fatal Four' were responsible for more than half (58.7%) the construction worker deaths in 2013, BLS reports. Eliminating the Fatal Four would save 468 workers' lives in America every year.

In 2012, the construction industry had 775 fatalities and the injury rate was 9.5%. The total number of fatal injuries in all the industries was 4,383. The construction industry in the United States accounted for 17.7%. Construction had the highest number of fatal injuries among all the industries in the United States in 2012. The national fatal injury rate on average was 3.2% (Bureau of Labor Statistics – US, 2014b).

There is no official statistics for the RMAA sector in the United States. The closet figure that reasonably represents RMAA works would be the statistics for the speciality trade contractors. According to the BLS (Bureau of Labor Statistics – US, 2015), the specialty trade contractor comprises 'establishments whose primary activity is performing specific activities (e.g., pouring concrete, site preparation, plumbing, painting, and electrical work) involved in building construction or other activities that are similar for all types of construction, but that are not responsible for the entire project'.

The work performed by speciality trade contractors may include new work, additions, alterations, maintenance and repairs. The production work performed by companies in this subsector is usually subcontracted from the general contractor or operative builders. For remodelling and repair construction works, they may be done directly for the owner of the property. Specialty trade contractors usually perform most of their work at the construction site, although they may have prefabrication and other work being done offsite on some occasions. Site preparation works for new construction are also included in this subsector (Bureau of Labor Statistics – US, 2015).

In 2013, specialty contractors' incidence rate of nonfatal occupational injuries was 4.2 per 100 workers whereas the average for the construction industry was 3.8 per 100 workers. Construction of buildings was 3.3 and heavy and civil engineering construction was 3.2. (Bureau of Labor Statistics – US, 2014a). With these figures, we can roughly conclude that the safety performance of the RMAA sector in the United States is worse than that of the national average.

Construction safety performance in the United Kingdom

According to Health and Safety Executive (2014), the construction industry accounted for only about 5% of the employees in the United Kingdom between 2013 and 2014. The construction industry in the United Kingdom accounted for 31% of fatal injuries to employees and 10% of reported major/specified injuries in the United Kingdom. Despite this, the safety performance of the construction industry in the United Kingdom has been improving. In 2013 to 2014, 42 fatalities occurred in the United Kingdom construction industry, of which 14 fatalities occurred to self-employed persons. The figures in 2013 to 2014 were somewhat better than the average of 46 over the previous 5 years, including an average of 17 to the self-employed. Besides physical injuries and fatalities, an estimated 2.3 million working days were lost in the year 2013/2014, including 1.7 million due to ill health and 592,000 due to workplace injury. These resulted in a total of 1.1 days lost per worker. Injuries and new cases of ill health of workers in the United Kingdom construction industry cost over GBP 1.1 billion (approximately USD 1.69 billion) a year (Health and Safety Executive, 2014).

Health and Safety Executive (2014) has attempted to compare safety performance of different sectors of the construction industry in the United Kingdom. Table 2.1 shows the percentages of volume of work categories in the UK construction industry in 2013/2014 and the respective percentages of accidents in the UK construction industry in 2013/2014. It is noted that due to different work category classification of the Office for National Statistics and the Reporting of Injuries, Diseases and Dangerous Occurrences (RIDDOR) required by Health and Safety Executive, the comparison has limitations. Work categories shown in Table 2.1 do not match perfectly with one another. Only a rough comparison of the volume of work and safety performance with reference to comparable work categories can be made. As indicated in Table 2.1, repair and maintenance accounted for about 36% of the construction output in 2013. However, it accounted for about 47% of the accidents in the construction industry. This indicates that repair and maintenance has higher risks than construction works (Health and Safety Executive, 2014).

Table 2.1 Comparing different sectors in the UK construction industry with their respective safety performance

Office for National Statistics categories	Percentage of volume of work in the UK construction industry in 2013/2014	RIDDOR work activities	Percentage of accidents in the UK construction industry in 2013/2014
New Housing & Other New Work	53%	New Building	23%
New Infrastructure	11%	Civil Engineering, Infrastructures, Roads, Bridges, Ports	10%
Repair and Maintenance	36%	Remodelling, Repairing, Extending, Building Maintenance	47%

Source: Health and Safety Executive (2014).

Construction safety performance in Hong Kong

The number of accidents in the construction industry of Hong Kong stood at 3,232 in 2013. Accident rate per 1,000 workers was 40.8 and the number of fatalities was 22 respectively in 2013 (Labour Department – Hong Kong, 2014). The construction industry in Hong Kong accounted for 22 out of 29 numbers (75.8%) of fatal industrial accidents in 2013.

The total number of construction accidents has been increasing. This may be due to more construction projects being rolled out. Table 2.2 shows that safety performance in terms of the accident rate per 1,000 workers of the construction industry has been improving. Accident rate per 1,000 workers fell from 60.3 in 2004 to 40.8 in 2013, representing a sizable decrease of 32.3%.

The same, however, does not apply to the RMAA sector. The percentages of the RMAA accidents in the construction industry (Table 2.2) have outweighed the percentages of the RMAA works in the construction industry since 2009 (Table 1.1). The percentages of RMAA accidents in the construction industry were 50.1 in 2009, 49.3 in 2010 and 44.7 in 2011 whereas the percentages of RMAA works in the construction industry were 48.2 in 2009, 45.0 in 2010 and 39.8 in 2011. The fatality rate of RMAA works in 2010 reached a new height at 66.7%. Six out of nine fatal accidents in the construction industry in 2010 were from the RMAA sector. With a disturbingly high accident and fatality rates in RMAA works, the need to improve safety performance in this growing sector is all the while more urgent.

Table 2.2 Industrial accidents of the construction industry

	2004	2005	2006	2007	2008	2009	2010	2011	2012	2013
(a) All reported construction accidents[*]	3,833 (17)	3,548 (25)	3,400 (16)	3,042 (19)	3,033 (20)	2,755 (19)	2,884 (9)	3,112 (23)	3,160 (24)	3,232 (22)
(b) Accident rate per 1,000 workers	60.3	59.9	64.3	60.6	61.4	54.6	52.1	49.7	44.3	40.8
(c) All reported accidents in RMAA works[*]	1,454 (6)	1,509 (12)	1,697 (9)	1,524 (6)	1,557 (8)[†]	1,379 (6)[†]	1,422 (6)[†]	1,390 (8)	N/A	N/A
Percentage of RMAA accidents to all reported construction accidents [(c)/(a)]	37.9%	42.5%	49.9%	50.1%	51.3%	50.1%	49.3%	44.7%	N/A	N/A
Percentage of fatal accidents in RMAA works to all fatal accidents in the construction industry	35.3%	48.0%	56.3%	31.6%	30.0%	47.4%	66.7%	34.8%	N/A	N/A

Note: Data from Labour Department of Hong Kong (2008, p. 3; 2014) and Legislative Council (2011a, 2011b).

[*]Figures in brackets denote the number of fatalities.

[†]Statistics for fatal RMAA accidents in 2008–2011 were collected from CIC (2013). N/A– Not available in public domain.

Sources: Construction Industry Council – Hong Kong (2013); Labour Department – Hong Kong (2008); Labour Department – Hong Kong (2014); Legislative Council – Hong Kong (2011); Legislative Council – Hong Kong (2009).

Safety regulations governing safety performance of RMAA works

Safety regulations in the United States

In the United States, the Occupational Safety and Health (OSH) Act is administered by the Occupational Safety and Health Administration (OSHA) (US Department of Labor, 2015). The OSH Act covers all industries and self-employed persons. Employers covered by the OSH Act must comply with the regulations and the safety and health standards stipulated by OSHA. Employers have a general duty under the OSH Act to provide a safe work environment to their employees. OSHA is responsible for safety inspections and investigations.

Part 1926 of Safety and Health Regulations for Construction stipulates the legislative requirement for employers to maintain safety and health of construction sites. According to the record-keeping requirement of OSHA, employers need to report to OSHA:

- any employee fatality as a result of a work-related incident;
- any in-patient hospitalization of one or more employees as a result of a work-related incident;
- any employee amputation as a result of a work-related incident;
- any employee loss of an eye as a result of a work-related incident.

Employers need to keep OSHA injury and illness records; however, companies with 10 or fewer employees are not required to routinely keep OSHA injury and illness records. In 2014, OSHA has released an updated list of almost 500 industry groups that are exempted from programmed safety inspections and the list includes five construction-related North American Industry Classification System (NAICS) codes. An employer would be exempted from a programmed inspection when there are 10 or fewer employees at a work site. The following are the five groups related to the construction industry:

- Power and communication line and related structures construction (NAICS 237130)
- Land subdivision (NAICS 237210)
- Other heavy and civil engineering construction (NAICS 237990)
- Electrical contractors and other wiring installation contractors (NAICS 238210)
- Other building equipment contractors (NAICS 238290)

In other words small RMAA contracting companies with 10 or fewer employees will be exempted from keeping injury and illness records. RMAA works involving small electrical/wiring installation/building equipment contractors with 10 or fewer employees will be exempted from programmed inspection.

Safety regulations in the United Kingdom

Health and Safety at Work etc. Act 1974

This Act requires employers to ensure the health and safety of their employees, other people at work and members of the public who may be affected by their work. There should be a health and safety policy. Employers who have five or more employees should have a written health and safety policy in place. For those who are self-employed, they should care for their own health and safety and make sure that their work would not endanger others. Employees should work together with the employer to uphold health and safety in the workplace and should not endanger themselves or put others at risk.

Management of Health and Safety at Work Regulations 1999

The Management of Health and Safety at Work Regulations 1999 (MHSWR) apply to everyone at work. They require employers to plan, control, organise, monitor and review employees' work. Employers are required to conduct risk assessment and identify strategies to mitigate any potential risks. Employers are also required to provide competent health and safety advice, information and training to their employees. Employers should put in place a plan to deal with serious and imminent danger and work with others who share the workplace to uphold safety and health.

Construction (Design and Management) Regulations 2015

The Construction (Design and Management) (CDM) Regulations 2015 (Health and Safety Executive, 2015) came into force in April 2015, replacing the Construction (Design and Management) (CDM) Regulations 2007. Same as the CDM Regulations 2007, the new one requires that health and safety is taken into account and managed throughout all stages of a project, from conception, design and planning through to site work and subsequent maintenance and repair of the structure.

Different from 2007 version, the CDM Regulations 2015 (Health and Safety Executive, 2015) apply to domestic clients, which are clients carrying out construction works not for the course of business. This means that small domestic extension and refurbishment works previously exempted are now included.

Another change is that CDM coordinator is not required anymore. Clients need to appoint a 'principal designer' on any project involving more than one contractor, which includes subcontractors. Most of the projects will need a principal designer.

Construction works which last for more than 30 working days and have more than 20 workers working at the same time or exceed 500 person days are required to notify the Health and Safety Executive before the commencement of the construction works.

Safety regulations in Hong Kong

Both new works and RMAA works contractors in Hong Kong are required to establish a company safety system which complies with the legislation and regulation governing safety of construction work. The regulations concerning safety of construction works are mainly Factory and Industrial Undertakings Ordinance, Construction Sites (Safety) Regulations Part VIII, and Occupational Health and Safety Ordinance (Construction Industry Institute – Hong Kong, 2007). Under the Factory and Industrial Undertakings Ordinance (Labour Department – Hong Kong, 2009), any person who works in a construction site needs to receive basic safety training to obtain a Construction Industry Safety Training Certificate (commonly known as a 'green card' in Hong Kong). Contractor has to employ a safety officer if there are 100 or more employees in one or more of the construction sites, and a safety supervisor in any site with 20 or more employees. Under the Construction Sites (Safety) Regulations Part VIII, all construction works, including repair and maintenance, need to inform the Labour Department before commencement. There is, however, an exemption from this regulation. Project with a duration of less than six weeks or fewer than ten workers are not required to inform the Labour Department (Hon, Chan, & Yam, 2012).

Under the Occupational Safety and Health Ordinance (Labour Department – Hong Kong, 2009), employers, occupiers and employees share the responsibility to ensure safety and health in their workplaces. Employers should ensure that plant and work systems are safe to operate. They should ensure plant or substances are safe to handle, store or transport. They should provide all necessary safety information, instruction, training and supervision. They should provide and maintain safe access to and egress from the workplaces and a safe and healthy work environment. Occupiers of premises should ensure that the premises, the means of access to and egress from the premises, and any plant or substance kept at the premises are safe and without risks to health to any person working on the premises, even if they do not directly employ that person on the premises. Employees are responsible for the safety and health of all persons at the workplace. They should follow the safety system or work practices set out by their employers. The Commissioner for Labour will issue improvement notices and suspension notices against workplace activity that may create an imminent hazard to employees. Contractors that fail to comply with these notices will be fined HKD 200,000 (approximately USD 25,641) and HKD 500,000 (approximately USD 64,103) respectively and imprisonment of up to 12 months (Hon et al., 2012).

Chapter summary

This chapter has introduced different safety performance measurement indicators in the construction industry. It has given a brief overview of safety performance of RMAA works in the United States, the United Kingdom and Hong Kong. Likewise, the safety regulations governing these jurisdictions have also been reviewed. Safety performance of the RMAA sector has revealed that this sector deserves more attention.

Accident data are useful information for industry practitioners. Brace, Gibb, Pendlebury, and Bust (2009) stated that 'a lack of proper accident data sets, whilst not an underlying cause in itself, could be a de-motivator and barrier to move things forward. Well compiled data, by contrast, will largely help the industry to inform intervention strategies as national averages and sweeping generalizations can lead people down the wrong path for their particular problems' (Brace et al., 2009).

In most jurisdictions, employment figures of the RMAA sector are not available. Thus, accident rate in terms of workers employed in the RMAA sector cannot be calculated, making cross-sector comparison in the construction industry impossible. Underreporting of minor injuries in RMAA projects is very common. It is also difficult for the government to monitor safety of RMAA projects because these projects are usually small in scale and have short project duration. Inadequate awareness of the importance of accident reporting and the traditional 'blame culture' within the construction industry also contribute to this disturbing situation (Chan et al., 2010).

Proper safety performance measurement indicators should be established. Comparable safety statistics should be compiled for this sector. This would require communication of government departments handling employment statistics and safety and health data respectively. While more concerted effort is needed to gather the employment figures in RMAA projects, more government inspections and law enforcement actions are needed to raise the overall level of accident reporting (Chan et al., 2010).

References

Australian Constructors Associations. (2015). *Lead Indicators: Safety Measurement in the Construction Industry*. Retrieved from www.constructors.com.au/wp-content/uploads/2015/09/Lead-Performance-Indicators-Guideline.pdf.

Biggs, H. C., Dingsdag, D. P., Kirk, P. J., & Cipolla, D. (2010). Safety culture research, lead indicators, and the development of safety effectiveness indicators in the construction sector. *International Journal of Technology, Knowledge and Society*, 6(3), 133–140.

Brace, C., Gibb, A., Pendlebury, M., & Bust, P. (2009). *Inquiry into the Underlying Causes of Construction Fatal Accidents. Phase 2 Report: Health and Safety in the Construction Industry: Underlying Causes of Construction Fatal Accidents – External Research*. Retrieved from www.hse.gov.uk/construction/ inquiry.htm.

Bureau of Labor Statistics – US. (2014a). Employer-reported Workplace Injuries and Illnesses- 2013. Bureau of Labor Statistics, US Department of Labor. Retrieved from www.bls.gov/news.release/archives/osh_12042014.pdf.

Bureau of Labor Statistics – US. (2014b). *National Census of Fatal Occupational Injuries in 2013*. Bureau of Labor Statistics, US Department of Labor. Retrieved from www.bls.gov/news.release/pdf/cfoi.pdf.

Bureau of Labor Statistics – US. (2015). *Industries at a Glance: Specialty Trade Contractors NAICS 238*. Bureau of Labor Statistics, US Department of Labor. Retrieved from www.bls.gov/iag/tgs/iag238.htm.

Census and Statistics Department – Hong Kong. (2009). *Report on the Quarterly Survey of Construction Output*. Retrieved from www.censtatd.gov.hk/products_and_services/products/publications/statistical_report/commerce_and_industry/index_cd_B1090002_dt_detail.jsp.

Chan, A. P. C., Wong, F. K. W., Yam, M. C. H., Chan, D. W. M., Hon, C. K. H., Wang, Y., Dingsdag, D., & Biggs, H. (2010). RMAA safety performance – how does it compare with Greenfield projects. In P. Barret, A. Dilanthi, R. Haigh, K. Keraminiyage & C. Pathirage (Eds.), *Proceedings of The CIB World Congress 2010, Building a Better World* (p. 154). Salford Quays, UK: CIB – International Council for Research and Innovation in Building and Construction.

Cheung, M. K. C. (2005). Speech by Mr Matthew CHEUNG Kin-chung, JP, Permanent Secretary for Economic Development and Labour at the Hong Kong Construction Association Annual Safety Conference on 30 August 2005. from www.labour.gov.hk/eng/major/300805.htm.

Construction Industry Institute – Hong Kong. (2007). *Construction Safety Involving Working at Height for Residential Building Repair and Maintenance: Research Summary*. Research Report No. 9, ISBN 978-988-99558-1-6.

Construction Industry Council – Hong Kong. (2013). *Hong Kong Construction Industry Performance Report for 2011. Version 1, April 2013*. Retrieved from https://www.hkcic.org/files/eng/documents/HKConstructionIndustry PerformanceReportfor2011(20130418)(English).pdf.

Department of Labor – US. (2015). *Occupational Safety and Health Administration – OSHA Law & Regulations*. Retrieved from www.osha.gov/law-regs.html.

Health and Safety Executive. (2006). *Health and Safety in Construction*. HSE, UK. Retrieved from www.hse.gov.uk/pubns/priced/hsg150.pdf.

Health and Safety Executive. (2014). *Health and Safety in Construction in Great Britain, 2014*. HSE, UK. Retrieved from www.hse.gov.uk/statistics/industry/construction/construction.pdf.

Health and Safety Executive. (2015). *Managing Health and Safety in Construction. Construction (Design and Management) Regulations 2015, Guidance on Regulations*. Retrieved from www.hse.gov.uk/pubns/priced/l153.pdf.

Hinze, J., Thurman, S., & Wehle, A. (2013). Leading indicators of construction safety performance. *Safety Science, 51*, 23–28.

Hon, C. K. H., Chan, A. P. C., & Yam, M. C. H. (2012). Empirical study to investigate the difficulties of implementing safety practices in the repair and maintenance sector in Hong Kong. *Journal of Construction Engineering and Management, 138*(7), 877–884.

Labour Department – Hong Kong. (2008). *Accidents in the Construction Industry of Hong Kong (1998–2007)*. Hong Kong: Labour Department, Government of Hong Kong. Retrieved from www.labour.gov.hk/eng/osh/pdf/Accidents ConstructionIndustry1998-2007.pdf.

Labour Department – Hong Kong. (2009). *Labour Legislation: Overview of Major Labour Legislation*. Retrieved from www.labour.gov.hk/eng/legislat/content4.htm.

Labour Department – Hong Kong. (2014). *Occupational Safety and Health Statistics Bulletin. Issue No. 14 (July 2014)* Occupational Safety and Health Branch, Labour Department, Government of Hong Kong. Retrieved from www.labour.gov.hk/eng/osh/content10.htm.

Legislative Council – Hong Kong. (2009) *Legislative Council Panel on Manpower – Hong Kong's Occupational Safety Performance in 2008. [LC Paper No. CB(2)2176/08-09(01)]*. Retrieved from www.legco.gov.hk/yr08-09/english/panels/mp/papers/mp0716cb2-2176-1-e.pdf.

Legislative Council – Hong Kong. (2011). *Legislative Council Panel on Manpower – Hong Kong's Occupational Safety Performance in the First Half of 2010 [LC Paper No. CB(2)814/10-11(04)]*. Retrieved from www.legco.gov.hk/yr10-11/english/panels/mp/papers/mp0120cb2-814-4-e.pdf.

Moore, J. T., Cigularov, K. P., Sampson, J. M., Rosecrance, J. C., & Chen, P. Y. (2013). Construction workers' reasons for not reporting work-related injuries: An exploratory study. *International Journal of Occupational Safety and Ergonomics, 19*(1), 97–105.

3 Accident causation models and safety management

Introduction

This chapter will provide a general review of accident causation models and different approaches to safety management. The main purpose of safety management is to prevent accidents from occurring. An effective safety management system has to take into account why accidents happened. Accident causation models are closely linked to safety management approaches. This chapter builds on the previous chapter of safety performance and provides the theoretical foundation of safety management.

Accident causation models

According to Toft, Dell, Klockner, and Hutton (2012), accident analysis models can be classified into simple linear models, complex linear models, and complex nonlinear models. The simple linear models perceive that accidents are caused by a series of linear events. Hence, accidents can be prevented by removing one of the sequential events in the chain. Complex linear models perceive that accidents are caused by a linear combination of unsafe acts and latent hazard conditions. Accidents can be prevented by strengthening barriers among the factors. Complex nonlinear models perceive that accidents are caused by a combination of interacting factors. Accidents can be prevented by understanding the complex relationships of interacting factors. Each type of model will be discussed below.

Simple linear model: Heinrich's domino model

According to Heinrich (1931), injury is the natural culmination of a series of events or circumstances. An accident is a link in the chain. Heinrich's domino model consists of five factors: social environment, fault of the person, unsafe acts, mechanical and physical hazards, accident, and injury (Toft et al., 2012). The model postulates that these accident factors line up like dominos. An accident happens when one of these factors falls and knocks down the rest of the factors. An accident is the result of the failure of the immediately preceding factor. Then injury is the result of an accident. Accidents can be prevented by removing one of the factors and stopping the knockdown effect (Toft et al., 2012).

Complex linear models

Energy-damage models

As cited in Toft et al. (2012), Viner (1991, p. 42) stated that 'damage (injury) is a result of an incident energy whose intensity at the point of contact with the recipient exceeds the damage threshold of the recipient'. The energy-damage model stipulates that a 'hazard is a source of potentially damaging energy and an accident, injury or damage may result from the loss of control of the energy when there is a failure of the hazard control mechanism' (Viner, 1991).

Time sequential model

The time sequential model is formed by Viner (1991). The model assumes accidents happen in the occurrence-consequence sequence. The model has three time zones. In Time Zone 1, there is 'some time between the event initiation and the event' (Toft et al., 2012), providing the opportunity to prevent an accident. In Time Zone 2, there is 'warning of the impending existence of an event mechanism' (Toft et al., 2012), providing the opportunity to take steps to reduce the likelihood of the event. Finally, in Time Zone 3, there is some time between when damage starts and when it completes, providing an opportunity to influence the outcome and the exposed groups (Viner, 1991).

Reason's Swiss Cheese model

Reason (1997) proffered the Swiss Cheese model. The model suggests that there are active errors and latent errors. The effect of the active errors can be seen immediately whereas that of the latent errors cannot be detected until they breach system defences. This model explains accident causes from human error to systems.

Complex nonlinear models

Systems-Theoretic Accident Model and Process (STAMP)

In systems theory, safety is defined as 'an emergent property that arises when the system components interact within an environment' (Leveson, 2004). Three basic concepts in STAMP are safety constraints, hierarchical safety control structures, and process models (Underwood & Waterson, 2013). Instead of events, causes of accidents in systems theory are the results of inadequate enforcement of constraints on behaviour at each level of a socio-technical system. Accidents occur when safety constraints are not enforced. Every hierarchical level of a system imposes constraints on and controls the behaviour of the level below it. Process models are an essential part of STAMP because they enable people to control the process effectively (Underwood & Waterson, 2013). Leveson (2011) stated that 'accident causation is a complex process involving the entire socio-technical system including legislators, government regulatory agencies, industry associations and insurance companies, company management, technical and engineering personnel, operators, etc.' Therefore, the accident process should be considered alongside proximate events in the event chain.

Functional Resonance Accident Model (FRAM)

According to Toft et al. (2012), FRAM was formed by Hollnagel (2004). FRAM is developed on the assumption that (Toft et al., 2012, p. 17)

> [an accident occurs in] highly complex and tightly coupled socio-technical systems in which people work. He describes systemic models as tightly coupled and the goals of organisations as moving from putting in place barriers and defences to focusing on systems able to monitor and control any variances, and perhaps by allowing the systems to be (human) error tolerant. ... FRAM is the first attempt to place accident modelling in a three-dimensional picture, moving away from the linear sequential models, recognising that 'forces (being humans, technology, latent conditions, barriers) do not simply combine linearly thereby leading to an incident or accident'. (Hollnagel, 2004, p. 171)

FRAM is founded on complex systemic accident theory and assumes that (Toft et al., 2012, p. 17)

> system variances and tolerances result in an accident when the system is unable to tolerate such variances in its normal operating mode. Safety system variance is recognised as normal within most systems, and represents the necessary variable performance needed for complex systems to operate, including limitations of design, imperfections of technology, work conditions and combinations of inputs which generally allowed the system to work. Humans and the social systems in which they work also represent variability in the system with particular emphasis on the human having to adjust and manage demands on time and efficiency. (Hollnagel, 2004, p. 168)

Accident analysis in construction

Studies in construction and engineering accident analysis tend to employ a linear approach (Love, Goh, & Smallwood, 2012) which is deemed inadequate for the design of appropriate preventive measures in the increasingly complex socio-technical environment of the construction industry. Traditional accident causation models, such as the Swiss Cheese Model and Normal Accident Theory model, assume an accident to be the result of a chain of events with direct cause-and-effect relationships, i.e., working back from the failure will find the root cause. Despite the inadequacies, these linear accident analysis models are still predominant in many construction companies.

These linear accident analysis models are popular in the construction industry because they seem to work well in analysing some accidents such as those arising from physical failure. Only when accidents arising from complex interaction of systems occur are linear models inadequate because they oversimplify causality and the accident process (Leveson, 2011). The linear accident analysis models fail to consider systemic factors leading to accidents, indirect or nonlinear interactions among events, or any causes of accidents lying among the interactions of well-functioning components (Leveson, 2011). Systems Theoretic Accident Modelling and Processes (STAMP) proffered by Leveson (2004, 2011) addresses this problem by applying systems theory to analyse accidents.

Traditional accident causation analysis in the construction industry has focused on the events underlying an accident, neglecting the accident process. Accidents are viewed as a chance occurrence of multiple independent events. Remedies have focused on eliminating the root cause event or adding barriers to stop propagation of failure events. Systemic factors, however, are not considered. For example, the government of the United Kingdom commissioned Brace, Gibb, Pendlebury, and Bust (2009) to conduct a thorough review of construction fatalities in the UK construction industry. Following the Swiss Cheese Model of Reason (1990), Brace et al. (2009) categorized the causes of fatal accidents into a number of macro- (e.g., society, education, industry, corporate organization, unions, Health and Safety Executive), mezzo- (e.g., project, management and organization, procurement) and micro-factors (e.g., worker, workplace, supervisor). When holes in the defensive plates are created by active or latent failures at each level and they line up, an accident trajectory will be formed. Based on the findings, the three major themes of preventive strategies recommended were 'culture and mindset', 'competency and training' and 'enforcement and compliance'. They all pointed to eliminating or adding barriers to accident-causing events.

Recently, Lingard, Cooke, and Gharaie (2013) analysed plant fatalities in Australia with reference to the accident analysis model developed by Loughborough University (Loughborough Model) (Health and Safety Executive, 2003). The Loughborough Model has the following three levels of incident causation: immediate circumstances, shaping factors and originating influences. Immediate circumstances are identifiable circumstances surrounding the accident. Shaping factors include supervision, site constraints, worksite design, communication, etc. Originating influences are client requirements, economic climate, construction education, etc. Although the Loughborough Model acknowledges that accidents originate from the interplay of different levels of factors, relationships between the different levels are still linear.

Systemic accident analysis provides a deeper understanding of how dynamic, complex system behaviour contributes to an accident (Underwood & Waterson, 2013). Under the theoretic concept of systems thinking, system dynamics is the 'methodology to understand the behaviour and dynamics of complex systems over time' (Goh & Love, 2012). Recently, system thinking has been applied in various arenas of construction safety. For example, a systems model of construction accident causation was formulated by Mitropoulos, Abdelhamid, and Howell (2005). Mohamed and Chinda (2011) used system dynamics modelling to study construction safety culture. More recently, Love, Lopez, Edwards, and Goh (2012) developed a systemic causal model of design error generation which is useful in developing a learning framework for design error prevention. It is anticipated that more and more application of systems approach will be seen in the analysis of construction accidents.

Accident causation models have become more and more complex. It does not mean that a more complex model is better than a simpler one. OSH professionals need to evaluate and select the most suitable one to adopt.

Evolution of safety management

Engineering approach

The earliest stage of safety management was an engineering approach. This period of time was often called the technology age. Safety was managed by focusing on technological improvement. At that time, most of the accidents were caused by technological failures such as machine breakdown. Safety was managed by ensuring the technology worked. Sophisticated risk assessment methods have been designed to ensure the equipment and technology was up to safety standard. According to Luria et al. (2008, p. 273),

> Safety engineering concentrates on safe physical environments including mechanical features for accident prevention and other features such as non-slip surfaces, railings, barriers for dangerous mechanical parts, noise insulation, and so forth. Under this approach dealing with safety issues is more of an engineering challenge than a managerial or behavioural concern.

According to Toft et al. (2012), this approach dominated until the Three Mile Island reactor failed. The accident of the Three Mile Island reactor triggered the safety experts to expand their thinking on how to manage safety. They realized that the accident was not caused by mechanical faults but by human error. Safety management should consider both the technology factor and the human factor.

Human error approach

The second stage of safety management is the human error approach. Safety experts realized that accidents would occur even when technology is safe. They started to focus on the interaction between technology and human behaviours. According to Health and Safety Executives (HSE, 2002, p. 38), the human behaviour factor contributes to approximately 80% to 90% of accidents.

According to Reason (2000), there are two aspects of human error problems, namely the person approach and the system approach. They result in different philosophies of error management. The person approach focuses on unsafe acts, that is, errors and procedural violations. Unsafe acts are perceived to be the result from peculiar mental processes such as forgetfulness, inattention, poor motivation, carelessness, negligence and recklessness. The pertinent preventive measures are mainly reducing unwanted variability in human behaviour: for example, poster campaigns appealing to one's sense of fear, disciplinary measures, threat of litigation, retraining, naming, blaming, and shaming.

For the system approach, humans are assumed to be fallible and errors are to be expected (Reason, 2000). Errors are seen as consequences rather than causes; for example, recurrent error traps in the workplace and the organisational processes that give rise to them. Preventive measures' target to change can change the work conditions rather than human action. The focus of the system approach is system defences. All hazardous technologies have barriers and safeguards. When an adverse event occurs, the important issue is not to find who committed the mistake, but how and why the defences failed.

Research studies on human errors have been done to try to better manage unsafe acts. Reason (2000) states that error management has two components which are limit the frequency of dangerous errors and create systems that can tolerate the occurrence of errors and hold their damaging effects. Supporters for the person approach would design strategies to help individuals be 'less fallible or wayward; adherents of the system approach and strive for a comprehensive management programme aimed at several different targets: the person, the team, the task, the workplace, and the institution as a whole.' Supporters of system approach will aim for high reliability organisations. High reliability organisations are systems operating in hazardous environments that have less than their proportional share of accidents, endure operational dangers, and achieve objectives of the organisations (Reason, 2000).

Organizational behaviour approach

Safety management in the past focused on job redesign, engineering approach and human factors to maintain workplace safety (Mullen, 2004). It is increasingly recognised that organizational factors should not be ignored in safety management. Hale and Hovden (1998) call this trend the third age of safety research. Organizational constructs (e.g., safety culture, safety climate, leadership and organizational commitment) are examined and relationships between organisational construct and safety performance (Johnson, 2007) are established.

Stages of safety management development have evolved from mechanistic, human error to a behavioural approach. Unsafe behaviour has been regarded as one of the main causes of accidents. As explained by Zohar (2002), contrary to the assumption that self-preservation overrides other motives (Maslow, 1970), people behave unsafely because safety measures are likely to entail modest apparent benefits but immediate costs, such as slower pace, extra effort or personal discomfort. If the likelihood of injury is underestimated in a seemingly safe environment, the expected utility of unsafe behaviour exceeds that of safe behaviour. Unsafe behaviour is reinforced as there is a tendency for people to place higher value on short-term results. In this sense, deterring unsafe behaviour is a great managerial challenge.

Unsafe behaviour is only the resultant symptom but there might be broader organisational and contextual factors that influence one's behaviour that are worth investigating. The behavioural approach suggests that there are reciprocal relations between attitudes, behaviour and situations (Bandura, 1986). Figure 3.1 illustrates their reciprocal relationships.

Safety climate, which is the perception of safety in the workplace, will change a person's cost-benefit perception of unsafe behaviour. Unsafe behaviours persist because they are naturally reinforced (Clarke, 2006). There are immediate benefits of taking shortcuts but punishment is rare or never happens. Positive safety climate offsets this natural reinforcement.

Organisational factors play an important role in improving safety performance. Organisational research studies on safety culture, safety climate, high performance work systems, safety leadership etc. proliferate.

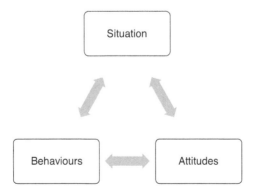

Figure 3.1 Reciprocal relationships between attitudes, behaviour and the situation.
Source: Adapted from Lingard and Rowlinson (2005), p. 319.

A high performance work system is a high-commitment- and high-involvement-oriented organizational strategy. Work practices generating high levels of commitment are found to increase safety behaviour. Zacharatos, Barling, and Iverson (2005) developed a model which shows interrelated relationships between high performance work system, trust in management, safety climate, safety incidents and personal safety orientation. A high performance work system exerts influence on safety incidents and personal safety orientation through the mediating variables 'trust in management' and 'safety climate'.

Safety leadership is another useful organisation construct to improve safety. In Lu and Yang (2010), safety leadership is defined by Wu (2005) as 'the process of interaction between leaders and followers, through which leaders can exert their influence on followers to achieve organizational safety goals under the circumstances of organizational and individual factors'.

Lu and Yang (2010) suggest that effective safety leadership helps to reduce the number of human errors and achieve a high level of safety. Research show that different styles of leadership have different influences on safety outcomes. Dimensions of transformational leadership have indirect effect on occupation safety (Barling, Loughlin, & Kelloway, 2002; Yule, Flin, & Murdy, 2007; Zohar, 2002). Transformational leadership shows greater concern for subordinates' welfare and develops closer individualized relationships, which promotes supervisory practices and in turn affects safety behaviour. Transformational leadership has a positive relationship with work group safety behaviour especially when the job's nature is not routine and safety procedures are not formalized. Transformational leadership allows people in the work group to make discretionary decisions while observing the general pattern of safety procedures. Transactional leadership is associated with higher accident rates (Barling et al., 2002). Corrective supervisors will not put safety into consideration if it is not one of the assigned priorities of the management (Zohar, 2002).

By addressing more fundamental organisational problems which adversely affect safety performance, many organizations experienced further breakthroughs and sustainable improvements of safety performance.

Systems approach

The systems approach is also called the holistic approach, which takes safety as an emergent property. This holistic approach aims to understand complex interrelationships and interdependencies between the technology, human and organisation (Toft et al., 2012). A holistic safety approach helps to explain safety from the complexities of the real world; for example, people use complex technology across multiple locations and divisions within the organisation to collaborate. The holistic approach provides a more complete or 'real' picture of the context. Thus, control measures and steps taken will be more efficient and effective in preventing accidents (Toft et al., 2012).

Safety management in construction

Technology

Recent technological advancement has begun to change safety management in construction. For example, Teizer, Allread, Fullerton, and Hinze (2010) developed a safety alert system with radio-sensing technology to warn workers when they stand too close to a dangerous zone. Cheng and Teizer (2013) made use of real-time three-dimensional data streaming to allow visualization of the construction site, which allows safety officers to monitor the construction project offsite. Radio Frequency Identification (RFID) technology has huge potential applications to improve construction safety management. Li et al. (2015) developed a real-time location system for tracking the location of workers to reduce unsafe behaviour and monitor safety management. Wider use of innovative technology, such as building information modelling, laser scanning, mobile computing and robotic vision is the impetus for a radical change in construction safety management. With leapfrog innovation in building information modelling (BIM) technology, virtual prototype construction has been widely applied to large construction projects (Ding, Wu, Li, Luo, & Zhou, 2011; Li, Chan, & Skitmore, 2012; Zhang & Fang, 2013). Simulating construction in virtual models enables the project participants to identify and fix the problems before commencement of the project (Ding et al., 2011; Li et al., 2012). Despite the positive impacts of innovative technology on safety management, there are hindrances as well. Tagging workers' location on site with wireless devices would bring about ethical and privacy concerns. Some studies show that workers being monitored more easily get depression, fatigue and strain.

Behaviour-based

Behaviour-based safety (BBS) 'focuses on what people do, analyses why they do it, and then applies a research-supported intervention strategy to improve what people do' (Geller, 2004). BBS emerges because there is a belief that most accidents are caused by unsafe behaviour. BBS focuses on providing stimuli to modify unsafe behaviour and reinforce safe behaviour (Choudhry, 2014). Behaviour is a function of consequences. Behaviour can be modified by increasing the cost of risky behaviour and increasing the benefits of being safe (Lingard & Rowlinson, 1994). This can be done by providing safety awards. Safety awarding is one of the most important safety strategies in construction. Safety awards given in the form of social reorganisation and praise was found to be the most effective (Hinze, 2006). Behaviour-based safety engages employees in the process of safety management, helping them to develop ownership to safety. The ultimate goal of behaviour-based safety is to change the safety culture.

Recently, there have been research initiatives to integrate BBS with technology. Li et al. (2015) developed proactive behaviour-based safety (PBBS), integrating BBS with the information technology called proactive construction management system (PCMS). PBBS provides a qualitative and quantitative way to improve construction safety. PBBS involves four steps, namely (1) baseline observation, (2) safety training, (3) follow-up observation, and (4) feedback and reinforcement. PCMS can gather real-time location-based behaviour data from workers on site. Supervisors can view and monitor site situation anytime offsite. This will lower safety management costs and improve efficiency. Through PCMS, workers will receive real-time warnings when they are exposed to risky situations and post real-time analyses when they may have behaved unsafely. PCMS can identify the location of workers, provide proactive warnings and monitor unsafe worker behaviours related to location. The PBBS approach gives objective data of workers' unsafe behaviours to safety supervisors/managers. Safety managers would have concrete evidence to show the workers about their unsafe behaviours and can increase the uptake of safer behaviours (Li et al., 2015).

Organizational behaviour

With the recognized importance of organizational constructs (e.g., culture, climate, leadership) in explaining causes of accidents, research has been done on safety culture, safety climate and safety leadership. Since safety culture and safety climate will be discussed in more detail in Chapter 7, this section focuses on safety leadership.

Workers were involved in unsafe behaviour because of a lack of safety awareness; to exhibit being 'tough guys'; work pressure; coworkers' attitudes; and other organizational, economic and psychological factors (Choudhry & Fang, 2008). However, there are underlying organizational factors contributing to their unsafe behaviours. The prompts for unsafe behaviour include a wide range of management system faults; for example, pressure to finish the job, achieving extreme construction targets, competing priorities, tight construction schedule, lack of training and lack of availability of equipment or materials (Choudhry, 2014).

Foremen or safety supervisors play an important role of safety management in construction sites (Kines et al., 2010). Dingsdag, Biggs, and Sheahan (2008) found that foremen/supervisors have a stronger influence on construction workers' safety attitudes than the workers' colleagues do. Zohar and Luria (2003) found that safety strategies targeting supervisors are more effective than those targeting employees. Besides the influence of supervisors, Lingard, Cooke, and Blismas (2011) found that the influence of coworkers on employees' safety behaviour should not be underestimated. Lingard, Cooke, and Blismas (2010) found that the injury rate of workgroups with a 'strongly supportive' group safety climate were only two-thirds of that of the workgroups with 'indifferent,' 'contradictory – mixed messages' or 'obstructive' group safety climates. Building strongly supportive group safety climate within the construction industry would likely reduce the number of injuries and improve safety.

Biggs, Dingsdag, and Roos (2008) developed 'A Practical Guide for Safety Leadership' to help the companies of the Australian construction industry to develop a safety competency framework and embed safety culture. Recommended steps include:

1 understand safety culture;
2 identify safety critical positions;
3 customise the task and position competency matrix;
4 adapt the competency specifications;
5 plan;
6 use a step-wise approach;
7 implement within your company.

Some large construction companies have already adopted and developed a safety competency framework, and the results are positive and encouraging. One company developed training programs by identifying current gaps between a construction safety competency framework and current organisational practice. Another company developed training programs by prioritising safety management tasks. One company developed training programs by mapping a construction safety competency framework and the Certificate IV in Business, majoring in occupational health and safety. Organizational strategies have been found to be useful in reducing injuries, removing unsafe behaviour and improving safety performance.

Chapter summary

This chapter briefly explained different accident causation models and approaches of safety management and their implications for construction safety. Although some of these approaches dominate at some point in time, they actually coexist. Accident causation models may be more applicable in certain industries and may remain prevalent over time. Having analysed why an accident happens, safety management is in place to provide appropriate strategies to prevent accidents. Multiple strategies may be adopted to tackle the safety problem from a combination of safety management approaches.

References

Bandura, A. (1986). *Social Foundations of Thought and Action: A Social Cognitive Theory*. Englewood Cliffs, N.J.: Prentice-Hall.
Barling, J., Loughlin, C., & Kelloway, E. K. (2002). Development and test of a model linking safety-specific transformational leadership and occupational safety. *Journal of Applied Psychology*, 87(3), 488–496.

Biggs, H. C., Dingsdag, D., & Roos, C. R. (2008). *A Practical Guide for Safety Leadership – Implementing a Safety Competency Framework.* Brisbane, Australia: Cooperative Research Centre for Construction Innovation.

Brace, C., Gibb, A., Pendlebury, M., & Bust, P. (2009). *Inquiry into the Underlying Causes of Construction Fatal Accidents. Phase 2 Report: Health and Safety in the Construction Industry: Underlying Causes of Construction Fatal Accidents – External Research.* Retrieved from www.hse.gov.uk/construction/inquiry.htm.

Cheng, T., & Teizer, J. (2013). Real-time resource location data collection and visualization technology for construction safety and activity monitoring applications. *Automation in Construction, 34*(9), 3–15.

Choudhry, R. M. (2014). Behavior-based safety on construction sites: A case study. *Accident Analysis & Prevention, 70,* 14–23.

Choudhry, R. M., & Fang, D. (2008). Why operatives engage in unsafe work behavior: Investigating factors on construction sites. *Safety Science, 46,* 566–584.

Clarke, S. (2006). The relationship between safety climate and safety performance: A meta-analytic review. *Journal of Occupational Health Psychology, 11*(4), 315–327.

Ding, L., Wu, X., Li, H., Luo, H., & Zhou, Y. (2011). Study on safety control for Wuhan metro construction in complex environments. *International Journal of Project Management, 29*(7), 797–807.

Dingsdag, D. P., Biggs, H. C., & Sheahan, V. L. (2008). Understanding and defining OH&S competency for construction site positions: Worker perceptions. *Safety Science, 46*(4), 619–633.

Geller, E. S. (2004, 01 OCT). Behavior-based safety: A solution to injury prevention: Behavior-based safety 'empowers' employees and addresses the dynamics of injury prevention. *Risk & Insurance, 15,* 66.

Goh, Y. M., & Love, P. E. D. (2012). Methodological application of system dynamics for evaluating traffic safety policy. *Safety Science, 50,* 1594–1605.

Hale, A. R., & Hovden, J. (1998). Management and culture: The third age of safety. A review of approaches to organizational aspects of safety, health, and environment. In A. M. a. W. Feyer, A. (Eds.), *Occupational Injury: Risk, Prevention and Intervention* (pp. 129–165). London: Taylor-Francis.

Health and Safety Executive. (2002). *Contract Research Report 430/2002: Strategies to Promote Safe Behavior as Part of a Health and Safety Management System.* HSE, UK. Retrieved from www.hse.gov.uk/research/crr_pdf/2002/crr02430.pdf.

Health and Safety Executive. (2003). *Causal factors in construction accidents. Research Report 156.* London.

Heinrich, H. W. (1931). *Industrial Accident Prevention: A Scientific Approach.* New York: McGraw-Hill.

Hinze, J. (2006). *Construction Safety* (2nd ed.). Gainesville, Florida: Alta Systems, Inc.

Hollnagel, E. (2004). *Barriers and Accident Prevention.* Aldershot, UK: Aldershot Ashgate.

Johnson, S. E. (2007). The predictive validity of safety climate. *Journal of Safety Research, 38,* 511–521.

Kines, P., Andersen, L. P., Spangenberg, S., Mikkelsen, K. L., Dyreborg, J., & Zohar, D. (2010). Improving construction site safety through leader-based verbal safety communication. *Journal of Safety Research, 41*(5), 399–406.

Leveson, N. (2004). A new accident model for engineering safer systems. *Safety Science, 42,* 237–270.

Leveson, N. (2011). Applying systems thinking to analyze and learn from events. *Safety Science, 49*(55–64).

Li, H., Chan, G., Huang, T., Skitmore, M., Tao, T. Y. E., Luo, E., Chung, J., Chan, X. S., & Li, Y. F. (2015). Chirp-spread-spectrum-based real time location system for construction safety management: A case study. *Automation in Construction*, *55*, 58–65.

Li, H., Chan, G., & Skitmore, M. (2012). Visualizing safety assessment by integrating the use of game technology. *Automation in Construction*, *22*, 498–505.

Lingard, H., Cooke, T., & Blismas, N. (2010). Properties of group safety climate in construction: The development and evaluation of a typology. *Construction Management and Economics*, *28*(10), 1099–1112.

Lingard, H., Cooke, T., & Blismas, N. (2011). Co-workers' response to occupational health and safety: An overlooked dimension of group-level safety climate in the construction industry. *Engineering, Construction and Architectural Management*, *18*(2), 159–175.

Lingard, H., Cooke, T., & Gharaie, E. (2013). The how and why of plant-related fatalities in the Australian construction industry. *Engineering, Construction and Architectural Management*, *20*(4).

Lingard, H., & Rowlinson, S. (1994). Construction site safety in Hong Kong. *Construction Management and Economics*, *12*(6), 501–510.

Love, P. E. D., Goh, Y. M., & Smallwood, J. (2012). Editorial. *Accident Analysis and Prevention*, *48*, 97–99.

Love, P. E. D., Lopez, R., Edwards, D. J., & Goh, Y. M. (2012). Error begat error: Design error analysis and prevention in social infrastructure projects. *Accident Analysis and Prevention*, *48*, 100–110.

Lu, C. S., & Yang, C. S. (2010). Safety leadership and safety behavior in container terminal operations. *Safety Science*, *48*(2), 123–134.

Luria, G., Zohar, D., & Erev, I. (2008). The effect of workers' visibility on effectiveness of intervention programs: Supervisory-based safety interventions. *Journal of Safety Research*, *39*(3), 273–280.

Maslow, A. (1970). *Motivation and Personality* (2nd ed.). New York: Harper and Row.

Mitropoulos, P., Abdelhamid, T. S., & Howell, G. A. (2005). Systems model of construction accident causation. *Journal of Construction Engineering and Management*, *131*(7), 816–825.

Mohamed, S., & Chinda, T. (2011). System dynamics modelling of construction safety culture. *Engineering, Construction and Architectural Management*, *18*(3), 266–281.

Mullen, J. (2004). Investigating factors that influence individual safety behavior at work. *Journal of Safety Research*, *35*, 275–285.

Reason, J. (1990). *Human Error*. Cambridge: Cambridge University Press.

Reason, J. (1997). *Managing the Risks of Organisational Accidents*. Aldershot: Ashgate.

Reason, J. (2000). Human error: Models and management. *British Medical Journal*, *320*(7237), 786–770.

Teizer, J., Allread, B. S., Fullerton, C. E., & Hinze, J. (2010). Autonomous pro-active real-time construction worker and equipment operator proximity safety alert system. *Automation in Construction*, *19*(5), 630–640.

Toft, Y., Dell, G., Klockner, K. K., & Hutton, A. (2012). Models of Causation: Safety. In Health and Safety Professionals Alliance (Ed.), *The Core Body of Knowledge for Generalist OHS Professionals*. Tullamarine, VIC: Safety Institute of Australia.

Underwood, P., & Waterson, P. (2013). Systemic accident analysis: Examining the gap between research and practice. *Accident Analysis and Prevention*, *55*, 154.

Viner, D. (1991). *Accident Analysis and Risk Control*. Melbourne: Derek Viner Pty Ltd.

Wu, T. C. (2005). The validity and reliability of safety leadership scale in universities of Taiwan. *International Journal of Technology and Engineering Education, 2*(1), 27–42.

Yule, S., Flin, R., & Murdy, A. (2007). The role of management and safety climate in preventing risk-taking at work. *International Journal of Risk Assessment and Management, 7*(2), 137–151.

Zacharatos, A., Barling, J., & Iverson, R. D. (2005). High-performance work systems and occupational safety. *Journal of Applied Psychology, 90*(1), 77–93.

Zhang, M., & Fang, D. (2013). A continuous behavior-based safety strategy for persistent safety improvement in construction industry. *Automation in Construction, 34*, 101–107.

Zohar, D. (2002). Modifying supervisory practices to improve subunit safety: A leadership-based intervention model. *Journal of Applied Psychology, 87*(1), 156–163.

Zohar, D., & Luria, G. (2003). The use of supervisory practices as leverage to improve safety behavior: A cross-level intervention model. *Journal of Safety Research, 34*(5), 567–577.

4   Fatal cases of RMAA works

Introduction

This chapter has analysed 146 fatal cases of RMAA works in Hong Kong which occurred between January 2000 and October 2014. Among the 146 fatal cases, 90 of them which occurred between 2000 and 2007 were collected from the Labour Department of the Hong Kong government. The remaining 56 occurred between January 2008 and October 2014 and were collected from WiseNews, a local newspaper archive in Hong Kong.

The fatal cases were analysed with a template jointly developed with the Labour Department of the Hong Kong government and the authors. Variables included in the analysis were time of accident, day of accident, month of accident, year of accident, type of accident, gender of victim, age of victim, trade of victim, length of experience of victim, body part injured of victim, injury nature of victim, place of accident, agent involved, type of work being performed, safety education and training of victim, use of safety equipment of victim, employment condition of victim, unsafe condition and unsafe action (Hon & Chan, 2013).

With the analysis of these 146 fatal cases of RMAA works in Hong Kong, a better understanding of the causes of accidents is made possible; and hence the formulation of preventive measures.

Analysis of fatal cases of RMAA works in Hong Kong

Time of accident

Two hours before and after lunch tend to have more RMAA fatal accidents. As shown in Figure 4.1, most of the fatal accidents occurred in the late morning and early afternoon, that is, between 10:01 and 12:00 (40 cases), and 14:01 and 16:00 (35 cases) respectively.

Day of week of accident

The beginning of the week tends to have greater numbers of accidents. As indicated in Figure 4.2, the greatest number of RMAA accidents in

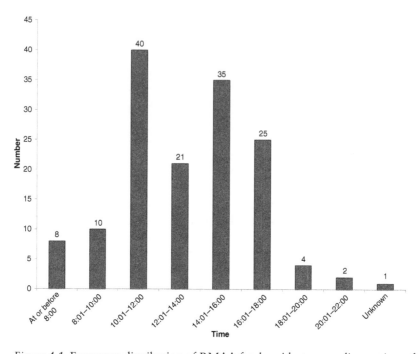

Figure 4.1 Frequency distribution of RMAA fatal accidents according to time of accident.

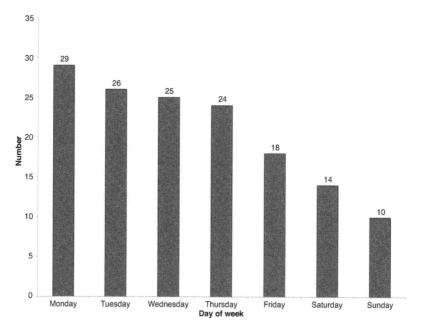

Figure 4.2 Frequency distribution of RMAA fatal accidents according to day of week.

Hong Kong between 2000 and 2014 occurred on Monday (29 cases). The number of fatal accidents declined towards the end of the week. Weekends have remarkably fewer occurrences of RMAA fatal accidents.

Month of accident

Summer tends to have more RMAA fatal accidents. Figure 4.3 shows that August had the largest number of RMAA fatal accidents in Hong Kong between January 2000 and October 2014. August is the mid-summer of Hong Kong and it is probably the hottest time of the year. The hot and humid climate in summertime of Hong Kong could easily lead to fatigue of workers. Workers are more susceptible to accidents in August.

Year of accident

With reference to Figure 4.4, the number of RMAA fatal accidents fluctuates over the years. It is likely to be linked to the volume of RMAA works in that year. It is likely that when there is a higher volume of RMAA works, the number of RMAA fatal accidents tends to increase.

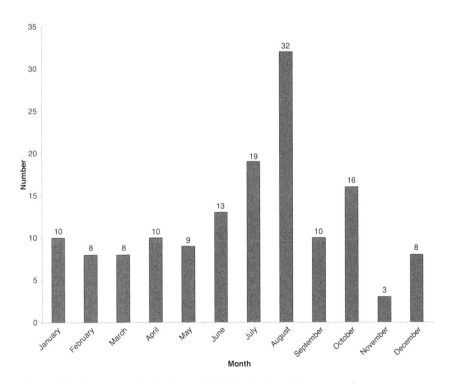

Figure 4.3 Frequency distribution of RMAA fatal accidents according to month.

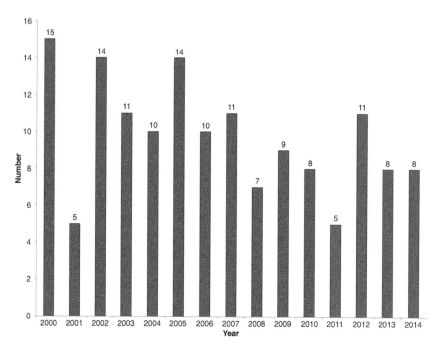

Figure 4.4 Frequency distribution of RMAA fatal accidents according to accident year.

Type of accident

Figure 4.5 shows that 'fall from height' is the top killer in the RMAA sector. Fall from height (96 cases) accounted for about 66% of the total RMAA fatal accidents between January 2000 and October 2014. The second greatest killer is contact with electricity or electrical charge (24 cases), which accounted for 16% of the total RMAA fatal accidents for the same period.

Gender of victim

Almost all victims of the fatal accidents of RMAA works in Hong Kong were male. During the period under scrutiny, there was only one female victim (Figure 4.6).

Age of victim

With reference to Figure 4.7, two age groups have notably high number of fatal cases. There were 25 victims aged between 30 and 34 and 23 victims aged between 55 and 59 who died in RMAA works.

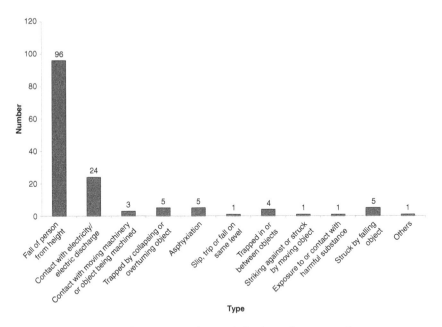

Figure 4.5 Frequency distribution of RMAA fatal accidents according to type.

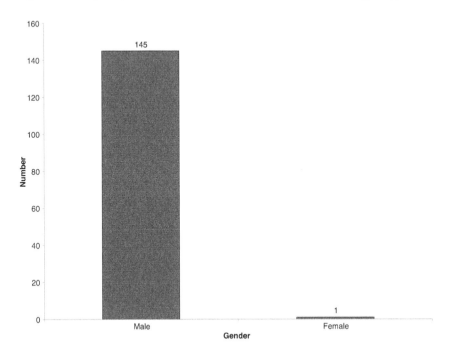

Figure 4.6 Frequency distribution of RMAA fatal accidents according to gender.

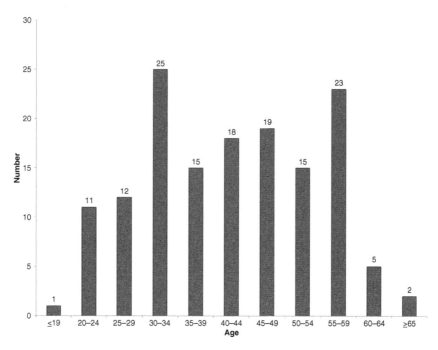

Figure 4.7 Frequency distribution of RMAA fatal accidents according to age.

Trade of worker

Apart from the category of 'others', a considerable number of victims (33 cases) were bamboo scaffolders (Figure 4.8).

Length of experience

Apart from those victims with unknown industry experience, those with 5 years or less of industry experience were more prone to fatal accidents. They were inexperienced and they were not able to foresee and assess potential risks involved in their work (Figure 4.9).

Body part injured

With reference to Figure 4.10, multiple locations and skull/scalp were the two most common body parts that the victims were injured.

Injury nature

Figure 4.11 shows that multiple injuries, contusion and bruise were the two most common injuries of the deceased.

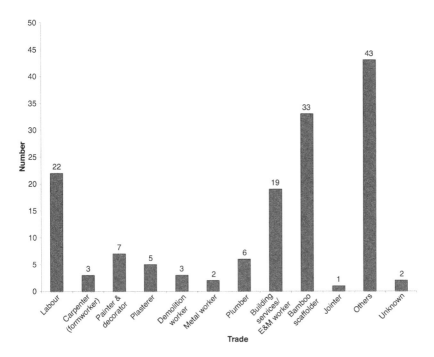

Figure 4.8 Frequency distribution of RMAA fatal accidents according to trade.

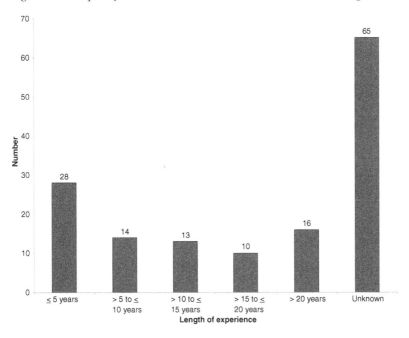

Figure 4.9 Frequency distribution of RMAA fatal accidents according to length of experience of the victim.

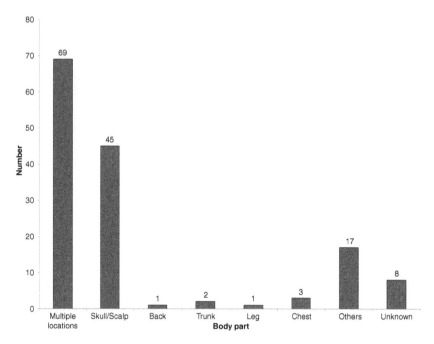

Figure 4.10 Frequency distribution of RMAA fatal accidents according to body part injured.

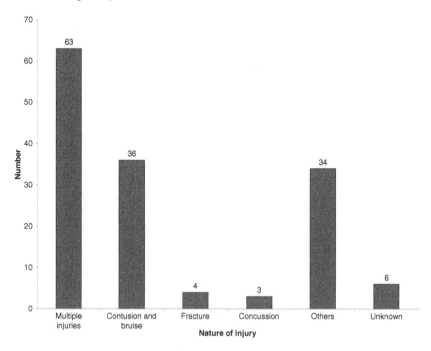

Figure 4.11 Frequency distribution of RMAA fatal accidents according to nature of injury.

Place of accident

Figure 4.12 indicates that most accidents occurred in external wall/facade and floor/floor opening.

Agent involved

As shown in Figure 4.13, apart from the group of 'others', scaffolding/gondola was the most often involved agent in RMAA fatal accidents in Hong Kong.

Type of work being performed

With reference to Figure 4.14, bamboo scaffolding (32 cases, 22%) was the most common type of work being performed by victims of the fatal accidents.

Safety education and training

Due to unavailability of information, most of the safety education and training experience of the victims was unknown. Even though Construction Industry Safety Training Certificate (commonly known as a Green Card) is mandatory for workers carrying out construction work in Hong Kong, it is surprising to find that eight victims did not have a Green Card (Figure 4.15).

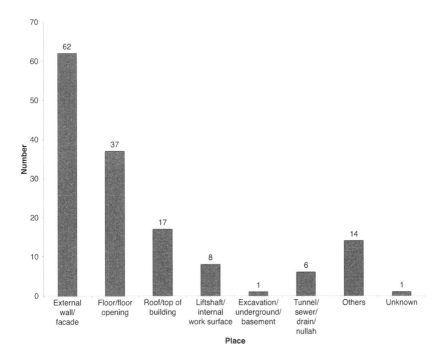

Figure 4.12 Frequency distribution of RMAA fatal accidents according to place.

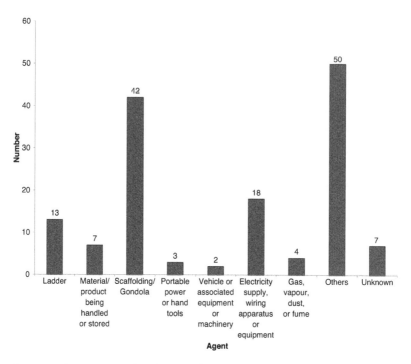

Figure 4.13 Frequency distribution of RMAA fatal accidents according to agent.

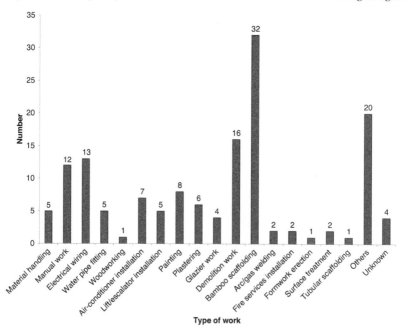

Figure 4.14 Frequency distribution of RMAA fatal accidents according to type of
work performed.

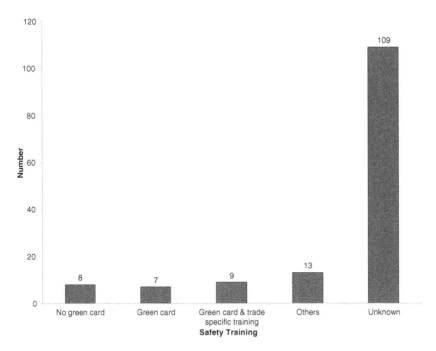

Figure 4.15 Frequency distribution of RMAA fatal accidents according to safety training.

Use of safety equipment

As shown in Figure 4.16, about 61 victims (42%) were not provided with safety equipment. It is also alarming that 20 victims (14%) died even when they were provided with and used safety equipment. This may suggest that the workers might not have used the safety equipment correctly or that substandard safety equipment was provided.

Employment condition

As shown in Figure 4.17, about 75% of the victims were employees. They largely rely on the employer to provide risk assessment, safety training and safety equipment.

Major categories of fall from height RMAA fatal accidents

Since fall from height was the top killer of the RMAA sector, further cluster analysis of 96 falls from height fatal cases was conducted to better reveal the characteristics of these fatalities. Cluster analysis is an exploratory analysis that tries to identify structures within the data. Objects with similar properties are grouped together. A two-step cluster analysis method which automatically determines the number of clusters was selected. Outlier treatment of noise handling was set at 25%. After forming the cluster features (CF)

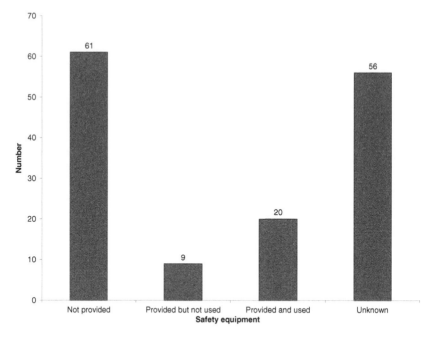

Figure 4.16 Frequency distribution of RMAA fatal accidents according to safety equipment.

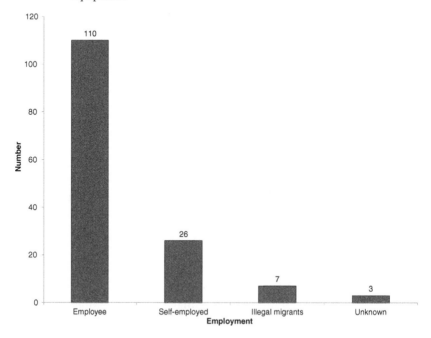

Figure 4.17 Frequency distribution of RMAA fatal accidents according to employment.

tree, outliers are either assigned to the CF tree or discarded, depending on their suitability to the CF tree (SPSS Inc., 2001). Six outliers were discarded. Three clusters encapsulating 90 fall from height RMAA fatal cases were formed (Figure 4.18). The value of 0.4 average silhouette of the cluster analysis indicates that the cluster quality is fair (below 0.2 is poor, 0.2–0.5 is fair, 0.5–1.0 is good) (SPSS Inc., 2001).

Referring to Figure 4.18, cluster 1 represents bamboo scaffolders who worked on bamboo scaffold/external facade and were aged between 35 and 39. Cluster 2 represents victims from other trades who died on floor/floor opening with other types of agents and were aged between 45 and 49. Cluster 3 represents building services/electrical and mechanical workers who died on ladders at external wall/facade and were aged between 55 and 59.

Table 4.1 shows analysis of key variables of fatal cases with reference to the three clusters. Most fall from height fatal accidents happened in the late afternoon (14:00–18:00). Fall from height accidents mainly occurred in the beginning and end of the weekdays. Summer is the month most prone to fall from height accidents. The age group 25–34 and 55 or older are more prone to fall from height. To be a bamboo scaffolder is relatively more dangerous than other types of trade. Many bamboo scaffolders were not provided with any safety equipment. Even if they used the safety equipment provided, 13 of them still died from fall from height.

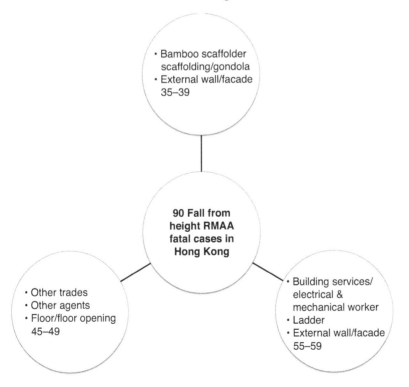

Figure 4.18 Cluster analysis of fall from height fatal cases in Hong Kong.

Table 4.1 Analysis of fall from height fatal cases in the RMAA sector of Hong Kong

Variable	Category	Cluster 1 (n = 40)	Cluster 2 (n = 26)	Cluster 3 (n = 24)	Total (N = 90)
Time	08:00–12:00	15	8	*10*	21
	12:01–14:00	6	5	1	11
	14:01–18:00	*19*	*10*	10	39
	Others	0	3	3	4
Day	In the beginning of weekdays	*14*	6	*11*	31
	In the middle of weekdays	8	3	4	15
	In the end of weekdays	11	*10*	7	28
	Weekends	7	7	2	16
Season	Spring	6	3	6	15
	Summer	*21*	6	*10*	37
	Autumn	6	9	4	19
	Winter	7	8	4	19
Age*	≤ 24	7	1	2	10
	25–34	*13*	1	2	16
	35–44	10	7	6	23
	45–54	3	*14*	4	21
	≥55	7	3	9	19
Trade*	Labour	2	10	1	13
	Painter and decorator	2	0	5	7
	Building services/ E&M worker	0	0	*10*	10
	Bamboo scaffolder	*29*	1	1	31
	Miscellaneous	7	*14*	7	28
Body part injured	Multiple locations	*28*	13	7	48
	Skull/scalp	8	13	*16*	37
Injury nature	Multiple injuries	*29*	*15*	8	52
	Contusion and bruise	6	9	*14*	29
	Fracture	2	1	3	6
	Others/unknown	3	0	2	5
Place*	External wall/ facade	*40*	3	12	55
	Floor/floor opening	0	*9*	*10*	19
	Roof/top of building	0	5	0	5
	Others	0	8	2	10

Variable	Category	Cluster 1 (n = 40)	Cluster 2 (n = 26)	Cluster 3 (n = 24)	Total (N = 90)
Agent*	Ladder	0	0	*13*	13
	Scaffolding/gondola	*38*	1	0	39
	Others	2	*25*	9	36
Type of work	Material handling	0	2	0	2
	Manual work	2	*6*	2	10
	Electrical wiring	0	1	*5*	6
	Water pipe fitting	0	0	*2*	2
	Woodworking	0	0	*1*	1
	Air-conditioner installation	0	1	*6*	7
	Painting	2	1	*5*	8
	Plastering	1	*2*	0	3
	Glazier work	1	*3*	0	4
	Demolition work	*4*	3	2	9
	Bamboo scaffolding	*28*	1	1	30
	Others	2	*6*	0	8
Safety equipment	Not provided	*20*	*15*	*11*	46
	Provided but not used	1	2	4	7
	Provided and used	13	4	0	17
	Unknown	6	5	9	20
Employment	Employee	*29*	*19*	*18*	66
	Self-employed	7	6	5	18
	Illegal migrants	2	1	1	4

Note: Numbers in bold and italics are the largest number in that variable according to cluster.

*Variables in cluster analysis.

As shown in Figure 4.19, unsafe process/job methods and improper procedures were the most frequent unsafe conditions leading to fall from height fatalities in the RMAA sector. For example, bamboo scaffolds were not fixed properly with three anchor bolts and safety belts were not tied to an independent lifeline.

With reference to Figure 4.20, failure to use safety belt/harness was the most frequent unsafe action leading to fall from height fatal accidents in the RMAA sector.

Lessons learned from case studies

Fall from height was the first killer in the RMAA sector. Bamboo scaffolder is the most dangerous trade in the RMAA sector. Substantial numbers of fall from height victims were young bamboo scaffolders. Many accidents occurred when they performed their work at an external wall/facade. 'Failure to use safety belt' was the major unsafe action of the victims who

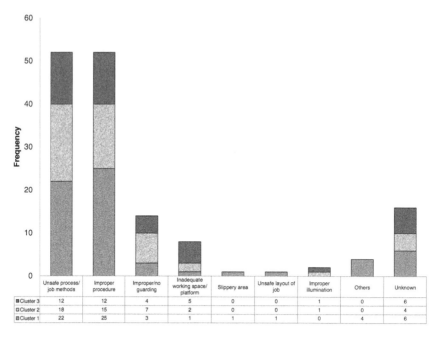

	Unsafe process/ job methods	Improper procedure	Improper/no guarding	Inadequate working space/ platform	Slippery area	Unsafe layout of job	Improper illumination	Others	Unknown
▣ Cluster 3	12	12	4	5	0	0	1	0	6
▢ Cluster 2	18	15	7	2	0	0	1	0	4
▣ Cluster 1	22	25	3	1	1	1	0	4	6

Figure 4.19 Frequency distribution of unsafe conditions leading to fall from height RMAA fatal accidents.

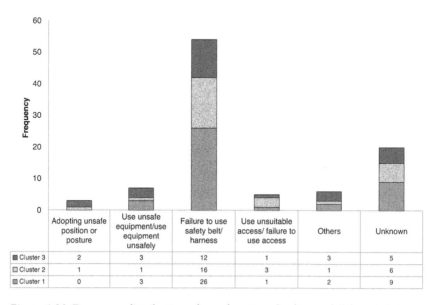

	Adopting unsafe position or posture	Use unsafe equipment/use equipment unsafely	Failure to use safety belt/ harness	Use unsuitable access/ failure to use access	Others	Unknown
▣ Cluster 3	2	3	12	1	3	5
▢ Cluster 2	1	1	16	3	1	6
▣ Cluster 1	0	3	26	1	2	9

Figure 4.20 Frequency distribution of unsafe actions leading to fall from height RMAA fatal accidents.

died from fall from height. Some victims died even with the use of safety belts because they did not attach their safety belts to independent lifelines or fixed anchor points. Safety belts had not been used properly. Young bamboo scaffolders were particularly susceptible to fatalities. RMAA works vary widely and it largely depends on the workers' experience to handle an ad hoc situation. Young and inexperienced bamboo scaffolders may not have assessed the risks involved fully when erecting or dismantling truss-out bamboo scaffold.

Many fatalities occurred when bamboo scaffolders worked on truss-out bamboos scaffold at an external wall/facade. Truss-out bamboo scaffold (as shown in Figure 4.21) is commonly used in Hong Kong for repair and maintenance works. The Buildings Department recommends that a truss-out bamboo scaffold should not exceed the height of 6 meters and the steel bracket underneath the truss-out scaffold should be fixed by three anchor bolts (refer to Figure 4.22). Sometimes, the bamboo scaffolders did not fully follow the requirements. Some old buildings do not have enough concrete strength to support the truss-out bamboo scaffold even when it is fixed with three anchor bolts. Truss-out bamboo scaffold is convenient and quick to erect; however, it is not safe to use truss-out bamboo scaffold to carry out repair and maintenance work at an external wall/facade.

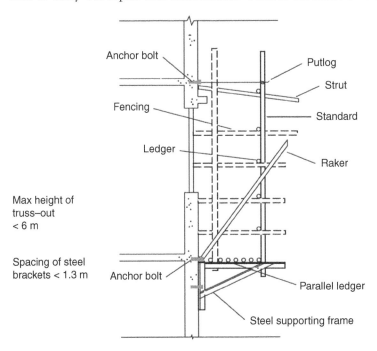

Figure 4.21 Truss-out bamboo scaffold.
Source: Buildings Department – Hong Kong (2006).

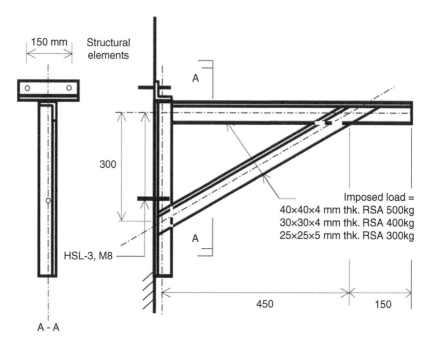

Figure 4.22 Base support for truss-out bamboo scaffold.
Source: Buildings Department – Hong Kong (2006).

A rapid demountable platform, which was designed to overcome the safety challenges of erecting and dismantling truss-out bamboo scaffold, will be discussed in Chapter 6.

The most accident-prone season is summer. The hot and humid summer is the peak season for repair and maintenance of air-conditioning systems, which largely involves work at height at an external wall/facade. Hot and humid weather may also affect the judgment of workers and lead to lapse of attention (Chan, Yam, Chung, & Yi, 2012). Heavy rain would affect the stability of truss-out bamboo scaffold. All these contribute to a higher accident rate in summer.

Many accidents occurred in the beginning of the week. Workers returning to work from the holiday may still have a holiday mood. Lapse of attention of workers working at height may have led to fatal accidents. Many RMAA workers were not provided with necessary safety equipment. Most of them did not wear a safety belt or harness. This reflects the inadequacy of RMAA workers' safety awareness. While the employer has a responsibility to provide safety equipment, RMAA workers should not perform any work at height without proper fall from height protection.

Chapter summary

This chapter analysed 146 fatal cases the RMAA sector in Hong Kong from January 2000 to October 2014. Fall from height was the top killer in the RMAA sector. Fall from height fatal cases were grouped into three clusters. The first cluster was bamboo scaffolders working at an external wall or facade. The second cluster was other trades of worker working on a floor or floor opening. The third cluster was building services workers working at the external wall or facade on a ladder. The detailed analysis of RMAA fatal cases has revealed characteristics of major types of accidents and provided lessons learnt to industry practitioners.

References

Buildings Department – Hong Kong (2006) Guidelines on the Design and Construction of Bamboo Scaffolds. Retrieved from www.bd.gov.hk/english/documents/code/GDCBS.pdf.

Chan, A. P. C., Yam, M. C. H., Chung, J. W. Y., & Yi, W. (2012). Developing a heat stress model for construction workers. *Journal of Facilities Management*, 10(1), 59–74.

Hon, C. K. H., & Chan, A. P. C. (2013). Fatalities of repair, maintenance, minor alteration, and addition works in Hong Kong. *Safety Science*, 51(1), 85–93.

SPSS Inc. (2001). *The SPSS Two Step Cluster Component: SPSS White Papers/ Technical Report TSCPWP-0101.* Chicago, IL: SPSS.

5 Safety problems and practices of RMAA works

Introduction

RMAA works have unique characteristics which render precarious safety problems. Safety practices appropriate for new construction works may not be appropriate for RMAA works. This chapter uses Hong Kong as an example to illustrate safety problems and practices of RMAA works in developed societies. Although Hong Kong is used as a case study, most of the problems and practices are common in developed societies. Causes of accidents and difficulties in implementing safety practices for RMAA works will be discussed. Discussion is based on research findings drawn from nine interviews with RMAA contractors and two rounds of Delphi survey with 13 multidisciplinary experts of RMAA safety in Hong Kong (Hon, Chan, & Wong, 2010). Details of the interview and Delphi survey can be found in Appendix A and B respectively.

Causes of RMAA accidents

RMAA works, being part of the construction industry, share some common safety problems with the construction industry. Due to unique characteristics of RMAA works, their causes of accidents are different from those of new construction works. Based on a study conducted in Hong Kong (Hon et al., 2010), there are mainly seven causes of RMAA accidents. They will be discussed below.

Low safety awareness of stakeholders in the RMAA sector

Risk-taking attitude of RMAA workers

Safety awareness of RMAA workers is influenced by the working environment they are in. RMAA works are often conducted within enclosed buildings or facilities. Their workplaces do not resemble construction sites, which are known to be dangerous. In fact, working in an enclosed workplace does not necessarily mean that it is not dangerous. On the contrary,

RMAA works frequently involve working at height, which imposes a high risk on workers. RMAA workers tend to underestimate dangers involved in their work because their working environment does not seem to be as dangerous as a new construction site.

One interviewee expressed that,

> workers do not perceive RMAA works as construction work. Although risk is the same, they tend to underestimate it. They have less safety awareness for RMAA works. In new works, they know very well that they have to wear safety helmets because they are within the area of construction site. However, in an occupied area, such as a theatre, they may not have the awareness to wear safety helmets. Actually, theatre has high headroom, working at height is also dangerous.

Safety awareness of RMAA workers is also influenced by the nature of tasks they perform. Very often, RMAA works involve small tasks that will last for a very short period of time. Considering that the tasks are small, they may find it troublesome to comply with safe practice. For minor RMAA work tasks, it is not surprising to find that the effort needed to follow safety practices would be larger than completing the task itself. The effort and time to be spent on safe practice does not seem to pay off. For convenience's sake, they may opt for taking risk instead of observing the safety practices.

The level of safety alertness of RMAA workers is often lower when they are dealing with minute tasks. Regarding this, one interviewee stated,

> Workers would have higher safety alert when they perform demolition works because demolition is known to be dangerous and the Buildings Department [the government department which regulates construction works in Hong Kong] poses strict control on demolition.

Safety awareness of RMAA workers is low because some of them are overconfident of their own expertise. For those who have been practised in the construction industry for a long time, they tend to believe that they have acquired the skills and expertise to accomplish the tasks. They may choose not to follow the recommended safe practices.

Less established contractors for RMAA works

Many contractors undertaking RMAA projects are small- and medium-sized companies. Safety monitoring systems of these companies may not be well developed or strictly enforced. Sometimes an RMAA main contractor sublets the whole project to subcontractors, who tend to be less knowledgeable about safety. Due to their small scale of business, they may not have the resources and expertise for upholding safety in their projects. With a small project size and value, it is also not economical to put in resources for safety. Subcontractors may not be able to provide full sets of personal protective equipment to workers and any safety supervision at all.

RMAA works involve mostly inexperienced clients

Property owners, as clients of RMAA works, may not have sufficient awareness for safety of the RMAA works undertaken in their properties. The most common practice is to employ a handyman without checking his qualification and safety record. Clients are not aware of the safety responsibility they have for the contractor or self-employed workers but only the cost. An owner's corporation may also go for the RMAA contractor with the lowest bid to undertake RMAA works, without considering the contractor's safety performance.

Inadequate safety supervision

RMAA jobsites are often dispersed. A typical example is a term maintenance contract, which requires the contractor to perform repair and maintenance services within a certain area or certain property developments of a client. For example, a contractor is solely responsible for repair and maintenance work of air-conditioning systems at all public housing estates in Kowloon, an urban area in Hong Kong. An RMAA contractor undertaking term maintenance contract of housing properties needs to send the team to perform maintenance work in different locations every day. Close safety supervision is difficult unless more resources are put into safety.

Many RMAA projects are small in scale and cannot afford to appoint a designated safety officer to take sole care of safety. Most RMAA projects will employ supervisors to look after both project supervision and site safety. Supervisors are usually experienced technicians with technical training primarily in construction technology and supervision, with some on-the-job training in site safety, whereas safety officers usually possess educational qualifications of safety education and training, such as a bachelor's degree in occupational safety and health. Safety officers are given the authority to stop unsafe work on site, whereas safety supervisors are not. Safety officers have to bear legal responsibility if they fail to perform their duty. For safety supervisors of RMAA works, safety is only one of the many tasks in their scope of work. They may tend to put safety at a lower priority than other project objectives when facing constraints. Similar to safety officers, safety supervisors may be required to submit a safety report to the management. However, it is unlikely that rectification of unsafe behaviour or condition is as immediate as for safety officers in new construction works.

Insufficient safety planning and hazard assessment

RMAA projects may face far more complex safety problems than new construction projects. Many types of RMAA works do not have standard method statements. Project method statements largely depend on the contractor's experience and project circumstances. For example, the concrete strength of an old building may have deteriorated because of ageing and improper maintenance. The project team needs to take this into account before removing or strengthening any part of the existing structure. Safety planning and hazard assessments may not be properly carried out because of time constraints and lack of experience.

Some RMAA works are undertaken in occupied buildings with all the utilities in use and occupants in place. Cutting off electricity for RMAA works is restrained. Workers may need to perform work without rendering dead the electrical system. With these constraints, safety planning and hazard assessment are often neglected. Besides, unforeseen safety problems may arise. All sorts of unforeseen hazards may occur in old developments; for example, the existing mechanical and electrical system is different from the as-built record, or illegal structures exist. Thus, it is difficult to standardize safety practice for RMAA works. Safe execution of RMAA works highly depend on the experience and self-motivation of workers and contractors.

Inadequate regulatory and monitoring system

Existing regulations and legislation tend to impose stringent requirement and monitoring on large projects. Small-scale projects may be exempted from the requirements and are not properly monitored. RMAA projects are often smaller in scale and thus exempted from the legislative requirements. For example, it is required by law that construction sites with more than 100 people should employ a safety officer. RMAA projects, which are often small in scale, are not required by law to appoint a full-time safety officer. Commencement of any construction projects with values greater than HKD 1,000,000 (approximately USD 128,000) should inform the Labour Department, which is the occupational safety and health regulatory body of the Hong Kong government. Many RMAA projects are less than this threshold and are exempted from reporting to the Labour Department. The Labour Department are not aware of these small projects until an accident happens, let alone monitor these RMAA projects while work proceeds.

Poor housekeeping and work environment

Poor housekeeping causes injury easily. RMAA work is often carried out in occupied buildings. Workers of different trades need to work in a limited space at the same time. For example, it is not uncommon to have over 100 workers working at the same time in a building space less than 2,000 m^2.

With multiple trades on site, materials and wastes may not be situated properly by different trades of workers. Unlike new construction works, RMAA project sites do not have enough designated space to store materials and accommodate wastes.

Besides, RMAA works conducted in enclosed building environments have poor ventilation. The work environment is congested, hot and stuffy. The adverse working environment makes RMAA workers not want to wear proper personal protective equipment unless doing so strictly enforced by the contractor or client.

Insufficient safety training of RMAA workers for handling multiple tasks

Very often, RMAA workers need to be multitasking. That is, RMAA works are not done by tradesman of a specific trade. For example, apart from painting, a painter may need to drill a hole in the wall, a skill which he is not familiar with, so as to complete his work. This would increase the chance of an accident. Unskilled workers often join the RMAA sector because they may think that trade skills of RMAA works are easy to pick up. Even though most RMAA workers have a Construction Industry Safety Training Certificate (known as a Green Card), they do not have specific safety training for handling multiple tasks of RMAA works.

Hurry to finish the work

Time is of the essence to every construction project. This is also true for RMAA works. RMAA works tend to have short project duration because of their small nature. Project duration is further compressed so as to reduce disturbances to existing occupants to the minimum. Sometimes RMAA works have to be conducted at night or over weekends when existing occupants have left. When workers are in a hurry to finish their tasks, they may cut corners and neglect safety.

As expressed by one interviewee, demolition tasks involved in RMAA projects are dangerous and need to be handled with care. However, workers of subcontractors being paid by a lump sum would like to finish the demolition tasks as fast as possible. Instead of knocking down a wall slowly piece by piece, they would choose to strike at the bottom to let the wall collapse quickly. They do this even they know that it is dangerous.

Relative importance of causes of RMAA accidents

Table 5.1 shows the relative importance of causes of RMAA accidents. 'Poor conscientiousness of RMAA works'; 'RMAA workers underestimate potential risks when performing small tasks for a short period of time'; and 'personal protective equipment not used, incorrectly used or not provided' were the top three important causes of RMAA accidents. All these indicate that safety awareness of RMAA workers is low. If proper personal protective equipment is not provided, safety awareness of RMAA contractors is low as well. The three least important causes of RMAA accidents ranked by the experts were 'low safety awareness of flat owners/tenants on RMAA works', 'inadequate regulatory control and monitoring system', and 'inadequate safety supervision'.

Table 5.1 Relative importance of causes of RMAA accidents

		Round 1		*Round 2*	
		Mean	*Rank*	*Mean*	*Rank*
1	Poor safety conscientiousness of RMAA workers.	4.46	1	4.54	1
2	RMAA workers underestimate potential risks when performing small tasks for a short period of time.	4.38	2	4.31	2
3	Inadequate safety supervision.	3.85	9	3.69	10
4	Low safety awareness of small/medium-sized contractors on RMAA works.	3.92	8	3.77	8
5	Low safety awareness of flat owners/tenants on RMAA works.	3.15	12	3.23	12
6	Inadequate site safety planning and hazard assessment.	4.00	6	4.00	7
7	Inadequate regulatory control and monitoring system.	3.38	11	3.46	11
8	Poor housekeeping and congested working environment.	3.62	10	3.77	8
9	Insufficient safety training of RMAA workers for handling multi-tasks.	4.15	4	4.08	6
10	Hurry to finish the work.	4.00	6	4.15	5
11	Lowest bid tendering method without pricing for safety items.	4.23	3	4.23	4
12	Personal protective equipment not used, incorrectly used or not provided.	4.08	5	4.31	2

Source: Hon, Chan, & Wong (2010).

It would be useful to compare these findings in Hong Kong with similar studies in other developed societies. Brace, Gibb, Pendlebury, and Bust (2009) is chosen for comparison because it is the latest comprehensive study of construction accidents commissioned by the government in the United Kingdom. RMAA accidents in the United Kingdom were included in this report.

Comparing the RMAA causes of accidents in Hong Kong with that of Brace et al. (2009) in the United Kingdom, they are quite similar with each other (Table 5.2). This probably indicates that causes of RMAA accidents in Hong Kong are chronic problems besetting the construction industry in other developed societies. Similar safety problems occur in different contexts of the construction industry. Characteristics of the RMAA sector tend to enlarge some of the common causes of accidents, resulting in more unsafe behaviours in the RMAA sector.

Low safety awareness of RMAA workers is the most important cause category of RMAA accidents. Originated from this category are 'Poor safety conscientiousness of RMAA workers'; 'RMAA workers underestimate potential risks when performing small tasks for a short period of time'; and 'personal protective equipment not used, incorrectly used or not provided'. It is particularly important for the workers in the RMAA sector to have good safety awareness because RMAA work is largely labour-intensive and does not rely much on machines.

After two rounds of Delphi exercises, an expert panel ranked 'poor safety conscientiousness of RMAA workers'; 'RMAA workers underestimate potential risk when performing small tasks for a short period of time'; and 'Personal protective equipment not used, incorrectly used or not provided' to be the three most important causes of RMAA accidents. These three are related to the workers. It seems that workers are to blame for the accidents having occurred. Actually, wider organizational factors such as job nature and pressure for production have to be considered. Some characteristics of RMAA work are conducive to unsafe behaviour of workers. Short project duration and minute tasks conducted in an occupied building give the wrong idea to workers that the job is not dangerous. These, coupled with difficulty of supervision in scattered locations, naturally reinforce unsafe behaviour of RMAA workers.

Table 5.2 Comparison of causes of accidents

Hon et al. (2010)	Brace et al. (2009)
• Low safety awareness of small-/medium-sized contractors on RMAA works.	• Immature corporate systems.
• Low safety awareness of flat owners/tenants on RMAA works.	• Nil.
• Inadequate regulatory control and monitoring system.	• Inappropriate enforcement.

Hon et al. (2010)	Brace et al. (2009)
• Lowest bid tendering method without pricing for safety items.	• Inappropriate procurement and supply chain arrangements.
• Poor safety conscientiousness of RMAA workers.	• Lack of individual competency and understanding of workers and supervisors.
• RMAA workers underestimate potential risks when performing small tasks for a short period of time.	• Poor behaviour.
• Inadequate safety supervision.	• Lack of ownership, engagement and empowerment of communication with, and responsibility for workers and supervisors.
• Inadequate site safety planning and hazard assessment.	• Site hazards.
• Poor housekeeping and congested working environment.	• Nil.
• Insufficient safety training of RMAA workers for handling multi-tasks.	• Ineffectiveness of lack of training and certification of competence.
• Hurry to finish the work.	• Poor employment practices.
• Personal protective equipment not used, incorrectly used or not provided.	• Poor equipment or misuse of equipment (including PPE).

Source: Hon, Chan, & Wong (2010).

As shown in Table 5.2, two causes of RMAA accidents—'low safety awareness of flat owners/tenants on RMAA works' and 'poor housekeeping and congested working environment'—found in Hong Kong were not found by Brace et al. (2009). These two causes typically reflect the features of RMAA works. Safety awareness of the general public is one of the causes of RMAA accidents identified by the interviewees, although it was ranked the lowest by the experts. 'General public' here includes homeowners, who are potential clients for small RMAA works. Homeowners, who are normally inexperienced clients, would simply hire handymen to conduct small RMAA work without considering the workers' safety capabilities and even qualifications. Poor housekeeping and congested working environment are particular problems of RMAA works. Different from new works, jobsites of RMAA works are confined by the building enclosures, so the jobsites have limited space for material storage. Different trades of workers come in the jobsite for a short time. Unless the RMAA contractor pays attention to housekeeping, the jobsite would be messy and easily lead to injuries.

Difficulties of implementing safety practices for RMAA works

Examining the difficulties of implementing safety practices in RMAA works will enhance our understanding for safety problems of this sector. Hon, Chan, & Yam (2012) conducted a comprehensive study to identify the difficulties of implementing safety practices for RMAA works which are summarised below.

Difficult to change the mindset of RMAA workers

Low safety awareness of workers is an industry-wide situation, especially for the RMAA sector. RMAA workers often perceive their tasks to be minute and easy to accomplish. Many of them take too lightly the risks involved in their work. For example, bamboo scaffolders commonly believe that it is safe to stand on an inner layer of bamboo scaffolds and alright for them not to wear a safety harness. Some workers use some fast but unsafe ways to move materials from place to place. Safety supervisors have to let the crew know what is not acceptable and check that the crew follows the instruction. In many cases, workers only pay lip service but they don't actually follow safety instructions.

Safety education and training could be the way to improve safety awareness. However, effective safety training is difficult for RMAA workers because the turnover rate for RMAA workers in a project is high. Different trades of RMAA workers may come in for a few days, then leave for another project. Currently, established RMAA contracting companies tend to select preferred subcontractors who share the same vision of safety as their priority. The main contractor would train workers of subcontractors as if they were their own direct labour. Well-trained direct labour of the main contracting company can act as a gauge for subcontractors' employees. They can also provide support for safety supervision by whistle-blowing or stopping unsafe behaviours. This arrangement is mutually beneficial. For the direct labour, they will benefit from better chances of promotion based on their good track records of performance. For the company, it can better monitor the safety performance of the subcontractors.

Most of the RMAA workers have some basic safety training and evidence (e.g., a Green Card) that they have completed a certified safety training course. However, Green Card certification has been criticized. There are different authorized bodies which can issue Green Cards. The quality of their training courses may vary and a standard is not properly monitored.

One interviewee echoed that

> most of the workers have done 32 hours of safety training but they
> don't treat it seriously. Instructors of these safety training courses are
> too lenient to workers and let them pass.

Difficult to supervise

RMAA works may disperse in different locations. One interviewee viv-
idly illustrated the difficulties of carrying out a term contract of repair and
maintenance:

> New works have morning safety briefing but it is not easy to implement
> in RMAA works. We tried to gather small groups of RMAA work-
> ers in different locations to have morning briefing sessions. For exam-
> ple, workgroup briefing sessions were held in five different locations in
> Sham Shui Po [a densely populated suburb in Hong Kong]. This strat-
> egy, however, was not very effective.

Even when sufficient personal safety equipment is provided, it is impossible to ensure every worker has used it. Unless more safety personnel are
employed, it is difficult for RMAA works to have the same level of safety
supervision as new works. A safety officer is not often employed in RMAA
projects. Instead, a site agent or foreman acts as a safety supervisor and
takes up the duty of safety supervision. Unlike a safety officer, who is solely
responsible for safety, safety is only part of the job duty for a supervisor.
Supervisors have less authority to stop unsafe behaviour and even if they do,
they may prioritise work progress if the schedule is tight. Safety of RMAA
works largely depends on whether supervisors strictly enforce safety regulations and good safety practices on site. If a supervisor imposes strict safety
supervision on the work crew, safety performance of that work crew is likely
to be better. If a supervisor gives lenient safety supervision to the work crew,
safety performance of that work crew is likely to be worse. However, supervisors often find it difficult to strike the right balance between progress and
safety.

Limited safety resources

Many RMAA works are conducted by small- and medium-sized subcontracting companies. They may lack resources for safety and may not be able
to afford to provide safety training and buy full sets of personal protective equipment for workers. For example, they may use old wooden ladders
instead of aluminium ladders. Old wooden ladders, having been used for

years, may be rotten and structurally unsound. The owners are also less aware of the importance of safety. Minor RMAA works are also small in scale which makes investing in safety seem less worthwhile.

One interviewee expressed that

> in recent years, the government of Hong Kong has reduced resources in safety. In the past, 2% of contract sum was provided in the contract for safety[1]. At present, the 2% of contract sum covers both safety and environmental protection. Resources for safety have been reduced... about 40%. Conditions of contract cannot reflect the safety needs of RMAA works. As more supervision is needed in RMAA works, the existing conditions of contract if used in RMAA work may need to be revised from assigning a safety supervisor for every 20 workers to every 10 workers.

Ad hoc site problems

Ad hoc site problems of RMAA projects make safety management difficult. The work activities of RMAA projects involve different risks from new construction projects. RMAA works are difficult to standardize, and the work environment of RMAA projects varies a lot. RMAA contractors face the challenge of providing sufficient information for workers to conduct their tasks safely.

Usually, RMAA contractors do not have complete control of the jobsite. A lot of coordination has to be done on site. RMAA works affect existing occupants and users of the building or even users of nearby buildings. Serious safety problem would occur if there is no proper project coordination. A typical case is that electricity supply needs to be shut down for repair and maintenance to be carried out. This needs coordination and liaison with different parties. Without effective coordination with the stakeholders, an electrician may be forced to undertake the work at risk.

Shortage of time

Time affects safety because when workers are in a hurry to finish their jobs, they are prone to neglect safety. In order not to disturb existing occupants, some RMAA works may only be able to be undertaken at night or over the weekends. If the RMAA works need to cut off electricity supply, a very rigid time limit would be allowed to carry out the work by negotiation with the property management company. Due to shortage of time, full implementation of safety practices is difficult. Shortage of time also affects proper safety planning and hazard assessment. RMAA works involve different trades of workers at one time. They may need to finish their part of work in a short period of time so as not to affect the work of other trades. Workers paid by a lump sum would also want to finish their job as quickly as possible so that they can move on to the next job. Safety is hindered by shortage of time.

Influx of illegal workers

RMAA works do not require heavy equipment, in most situations, and need only to have handyman tools. Illegal immigrants may find RMAA a sector relatively easy to enter. They may be introduced to work in RMAA projects by their friends who work for RMAA subcontractors. However, they are not protected by law to undertake RMAA works and they may not be conscious of the dangers pertinent in RMAA works because they are not familiar with safe practice in Hong Kong, and they may not have the necessary skills to perform the tasks safely.

Relative importance of difficulties of implementing safety practices in RMAA works

The relative importance of difficulties of implementing safety practices in RMAA works are summarised in Table 5.3.

The top three important difficulties of implementing safety practices in RMAA works are 'limited safety resources for RMAA projects undertaken by small/medium-sized contractors', 'difficult to change the mindset of RMAA workers' and 'difficult to conduct safety supervision due to scattered locations'.

Table 5.3 Relative importance of difficulties of implementing safety practices in RMAA works

		Round 1		Round 2	
		Mean	Rank	Mean	Rank
1	Difficult to change the mindset of RMAA workers.	3.54	4	3.92	2
2	Difficult to conduct safety supervision due to scattered locations.	3.69	2	3.77	3
3	Limited safety resources for RMAA projects undertaken by small-/medium-sized contractors.	4.15	1	4.08	1
4	Difficult to standardize the operational procedures of RMAA works due to ad hoc site problems.	3.00	8	3.08	8
5	Shortage of time to deal with safety issues.	3.08	7	3.23	7
6	High turnover rate of RMAA workers.	3.54	4	3.54	5
7	Small scale and short duration of RMAA projects.	3.62	3	3.62	4
8	Influx of illegal workers.	2.85	9	2.85	9
9	Difficult to control self-employed workers.	3.54	4	3.54	5

Safety performance of small- and medium-sized companies has been identified as an area for further investigation (Hinze & Gambatese, 2003). Small- and medium-sized companies often find themselves isolated from the latest safety and health information because they are difficult for safety and health regulatory bodies to reach (Halse, Bager, & Granerud, 2010). Limited resources of these companies is the major hindrance for improving their safety performance. Another challenge for implementing safety practices in RMAA sector boils down to the safety awareness of workers and their willingness to comply with safety practices. Scattered location of RMAA works further increases the implementation challenges. Difficulties for implementing safety practices of RMAA works are signposts of future safety strategies. If they are properly tackled, safety practices in the RMAA sector can be better implemented and thus further improve safety performance of the RMAA sector and the construction industry as a whole.

Chapter summary

This chapter summarises safety problems of RMAA works in Hong Kong. Causes of RMAA accidents and difficulties of implementing safety practices of RMAA works in Hong Kong have been discussed.

Safety awareness of RMAA workers, contractors and clients is insufficient. Safety awareness of RMAA workers is limited by the work environment and minute tasks involved whereas that of contractors is limited by the small scale of the business and project value. Safety awareness of clients is limited by their experience and sole consideration of cost. The mindset of workers is difficult to change because of their strong belief in their expertise, seemingly safe work environment and high turnover rate.

Safety supervision is inadequate and difficult because RMAA works tend to be dispersed in location. This is particularly the case for term maintenance contracts. Safety supervisors cannot give close supervision to project sites which are widely dispersed.

Resources for safety are often limited in RMAA projects. It is unlikely to be cost-effective for small-sized RMAA contractors to invest in safety for projects with a small contract sum.

Safety planning and hazard assessment is insufficient because RMAA works tend not to have a standard method statement to follow but depend on the experience of the contractor and site team. Ad hoc problems often arise.

Monitoring and regulatory systems for RMAA works are inadequate. Small-sized RMAA projects are exempted from some existing legal requirements; for example, no safety officer is required for small projects and no notification about commencement of the project to the Labour Department is needed.

Poor housekeeping and work environment lead to RMAA accidents. RMAA project sites are small and congested. Storage area for material and waste is often limited. Working in enclosed building is hot and stuffy, making the workers unwilling to wear personal protective equipment.

Training for multiple tasks is lacking. RMAA workers may not be aware of hazards involved in tasks that they are not familiar with. Currently, there is no safety training tailor-made for RMAA works.

Shortage of time is a particular problem for RMAA works. In order not to disturb users of the existing facility, RMAA works need to be conducted at night or over weekends. When workers are in a hurry to finish their work, they often neglect safety.

Hinze (2008) states that construction safety problems rarely exist in only a single country. Instead, construction industries of different jurisdictions face similar safety problems. Although safety problems of RMAA works are discussed based on findings in Hong Kong, they can be extrapolated to be in other developed societies with an expanding RMAA sector and emerging safety problems.

Note

1   This is a special clause of Pay for Safety and Environment Scheme (PFSES) in Hong Kong. For details of PFSES, please refer to Chapter 9.

References

Brace, C., Gibb, A., Pendlebury, M., & Bust, P. (2009). *Inquiry into the Underlying Causes of Construction Fatal Accidents. Phase 2 Report: Health and Safety in the Construction Industry: Underlying Causes of Construction Fatal Accidents – External Research*. Retrieved from www.hse.gov.uk/construction/inquiry.htm.

Halse, P., Bager, B., & Granerud, L. (2010). Small enterprises – Accountants as occupational health and safety intermediates. *Safety Science, 48*(3), 404–409.

Hinze, J. (2008). Construction safety. *Safety Science, 46*(4), 565.

Hinze, J., & Gambatese, J. (2003). Factors that influences safety performance of speciality contractors. *Journal of Construction Engineering and Management, 129*(2), 159–164.

Hon, C. K. H., Chan, A. P. C., & Wong, F. K. W. (2010). An empirical study on causes of accidents of repair, maintenance, minor alteration and addition works in Hong Kong. *Safety Science, 48*(7), 894–901.

Hon, C. K. H., Chan, A. P. C., & Yam, M. C. H. (2012). Empirical study to investigate the difficulties of implementing safety practices in the repair and maintenance sector in Hong Kong. *Journal of Construction Engineering and Management, 138*(7), 877–884.

6 Engineering solutions to RMAA works

Introduction

This chapter introduces different engineering devices which help improve the safety of RMAA works. These include rapid demountable platform (RDP), developed by a research team led by the first author of this book, and alternative devices found in the industry. Finally, this chapter will also mention prevention through design.

Rapid demountable platform (RDP)

Background

As discussed in Chapter 4, bamboo truss-out scaffold, which is commonly found in RMAA works of Hong Kong, has caused many deaths. Bamboo truss-out scaffold is often necessary because it is not economical to build a scaffold tower which rests on the ground to reach a flat in a high-rise building for minor RMAA works; for example, changing and cleaning of air- conditioners. Fatalities have been caused by the collapse of a bamboo truss-out scaffold or workers climbing out of a flat to install or dismantle a bamboo truss-out scaffold. Many accidents actually happen during the installation or removal of bamboo truss-out scaffold rather than when using it.

Construction Industry Institute – Hong Kong (2009) identified a number of safety problems commonly found in bamboo truss-out scaffold. Bamboo truss-out scaffold should be supported by brackets underneath (see Figures 4.21 and 4.22). Each of these brackets should be anchored to the wall with three anchor bolts. Low-quality anchor bolts are often used instead, which may not provide sufficient support to the bracket. Worse still, the third anchor bolt is usually difficult to affix properly. For the steel bracket, there are no required standards, and those steel brackets self-made by workers may not be strong enough. Conditions of the external wall of RMAA works are unpredictable. Workers might not use any safety equipment when erecting or dismantling bamboo trust-out scaffold. In light of these problems, rapid demountable platform (RDP) was developed with a hope to replace or supplement existing bamboo truss-out scaffold.

Design principles

Development of RDP has two stages. The first stage, RDP-1, aims to develop a working platform which does not require usage of steel brackets and anchor bolts and can be fixed from inside of the flat. The first-stage design was made of steel and wood. RDP-1 was released and evaluated by industry practitioners. The design of RDP-1 was then further improved by the research team in the Hong Kong Polytechnic University to develop RDP-2. The second stage, RDP-2, was to improve the original design with specific consideration to make use of lightweight material, improve the package for better mobility, control the production cost and apply to RMAA works in Hong Kong (Construction Industry Institute – Hong Kong, 2009).

RDP-2 is a modular design with two design philosophies. First, most of the parts with the same function are standard in colour, form and dimension. Second, the design is user-friendly and safe by incorporating product semantics and an innovative interlocking system. As shown in Figure 6.1, toe-boards are in place to reduce the chance of falling objects.

A metallic colour scheme with bright and florescent colour and zebra hatching patterns were added on some parts of the RDP to provide warning to workers. The metallic colour scheme is adopted for the main components and the bright red colour of the triangular frame signifies the strength of the

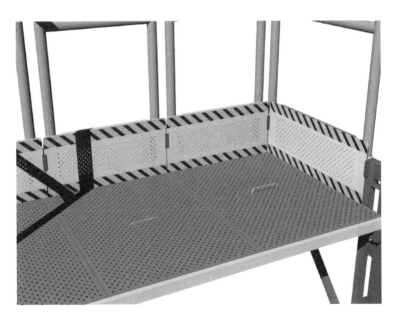

Figure 6.1 The platform is surrounded by toe-boards to avoid falling objects.
Source: Construction Industry Institute – Hong Kong (2009).

Figure 6.2 Metallic strips are added to the toe-boards.
Source: Construction Industry Institute – Hong Kong (2009).

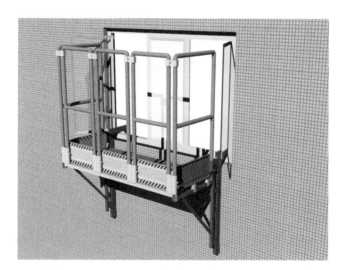

Figure 6.3 RDP with metallic strips.
Source: Construction Industry Institute – Hong Kong (2009).

Figure 6.4 A new interlocking system was designed to allow flexible joining with horizontal rotation.
Source: Construction Industry Institute – Hong Kong (2009).

system. The eye-catching florescent colour and the hatching lines/patterns in the toe-boards serve as a warning sign to the passersby (Figures 6.2 and 6.3).

Another feature of RDP is the interlocking system (Figure 6.4). The interlocking system is installed to the railings to allow flexible joining of railings with horizontal rotation.

Key components

RDP-2 consists of four components. They are as follows:

* Supporting frame unit (SFU) (Figure 6.5);
* Triangular frame unit (TFU) (Figure 6.6);
* Rail panel and toe-board unit (RTU) (Figure 6.7);
* Platform panels unit (PPU) (Figure 6.8).

Figure 6.5 Supporting frame unit (SFU).
Source: Construction Industry Institute – Hong Kong (2009).

Figure 6.6 Triangular frame unit (TFU).
Source: Construction Industry Institute – Hong Kong (2009).

Figure 6.7 Railing panels and toe-boards unit (RTU).
Source: Construction Industry Institute – Hong Kong (2009).

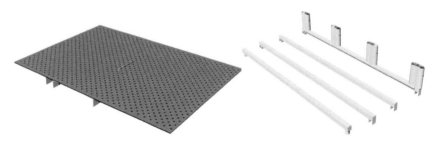

Figure 6.8 Platform panels unit (PPU).
Source: Construction Industry Institute – Hong Kong (2009).

To strike a balance between safety and weight, components of RDP-2 are made of a combination of stainless steel and aluminium alloy. The railings are made of stainless steel pipes and the platform panels and toe-boards are aluminium alloy. Total weight of the RDP is approximately 81 kg.

Installation procedures

Installation of RDP is simple. It involves four steps:

*   Step 1: Insert the SFU to the parapet wall through the window frame (Figure 6.9). Adjust the height of the SFU to an appropriate level (Figure 6.10). Tighten the screws of SFU to the wall.
*   Step 2: Install the TFU to the SFU at the desired level. Insert the anchor pin into the slot at the top of the TFU and SFU (Figures 6.11 and 6.12).
*   Step 3: Install the PPU to the TFU. Slot each set of RTU to the SFU and secure by the pin on the socket (Figures 6.13 and 6.14).

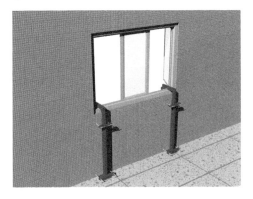

Figure 6.9 Insert the SFU to the wall.
Source: Construction Industry Institute – Hong Kong (2009).

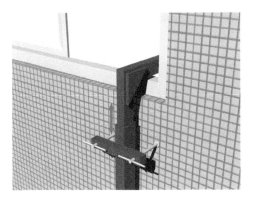

Figure 6.10 Adjust the height of the SFU.
Source: Construction Industry Institute – Hong Kong (2009).

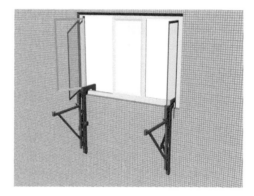

Figure 6.11 Install the TFU to SFU.
Source: Construction Industry Institute – Hong Kong (2009).

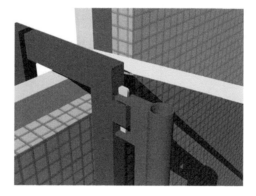

Figure 6.12 Insert the anchor pin.
Source: Construction Industry Institute – Hong Kong (2009).

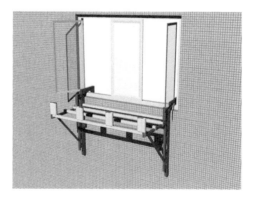

Figure 6.13 Install the PPU to the TFU.
Source: Construction Industry Institute – Hong Kong (2009).

- Step 4: Interlock the RTU by the preinstalled locking system on the railing panels. Check the tightness of the screws of the SFUs (Figures 6.15 and 6.16).

A completely installed RDP is shown in Figure 6.17 and Figure 6.18. For dismantling the RDP, order of these steps should be reversed (Construction Industry Institute – Hong Kong, 2009).

Figure 6.14 Secure the PPU and TRU by pin.
Source: Construction Industry Institute – Hong Kong (2009).

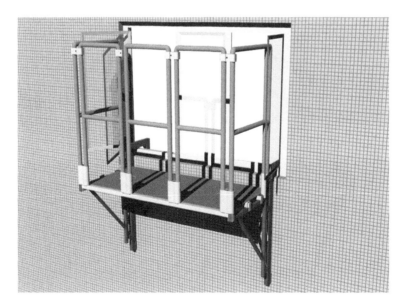

Figure 6.15 Install railings to the PPU.
Source: Construction Industry Institute – Hong Kong (2009).

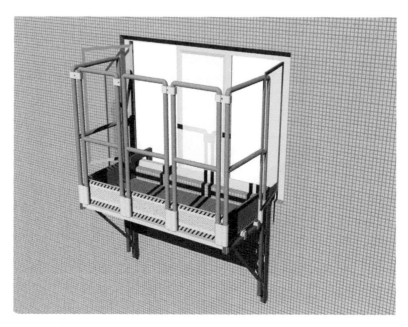

Figure 6.16 Install toeboard to the PPU.
Source: Construction Industry Institute – Hong Kong (2009).

Figure 6.17 Underneath of RDP.
Source: Construction Industry Institute – Hong Kong (2009).

Figure 6.18 Elevation of RDP.
Source: Construction Industry Institute – Hong Kong (2009).

Advantages of RDP

Performance of RDP and truss-out bamboo scaffold is compared in Table 6.1. RDP outperforms truss-out bamboo scaffold in every aspect, although it requires a bit higher initial cost.

In terms of cost, each RDP costs about HKD 6,000 (approximately USD 769) to HKD 10,000 (approximately USD 1,282) whereas erecting a bamboo truss-out scaffold costs about HKD 2,000 (approximately USD 256) to HKD 3,000 (approximately USD 385) each time. RDP lasts for at least 5 years if properly maintained. In that case, RDP costs only about HKD 2,000 (approximately USD 256) each year (Cheung & Chan, 2012).

In terms of duration of installation, RDP takes 10 minutes to erect or dismantle whereas bamboo truss-out scaffold takes 3 hours to erect and 2 hours to dismantle. RDP allows workers to dismantle it every day without leaving it at a window. In contrast, truss-out scaffold has to remain until the repair and maintenance work finishes. Day-to-day installation and removal of RDP reduces the likelihood of potential break-in after work. (Cheung & Chan, 2012)

In terms of application, RDP perfectly fits for a variety of jobs such as external building inspection, change of air-conditioning unit, maintenance on plumbing/drainage system, fixing of water seepage, painting, plastering and tiling/rendering of external walls. The only limitation of RDP is that it is designed to fit most of the windows at residential buildings but the current design may not suit planter or bay windows. It is not as flexible as truss-out bamboo scaffold to reach around corners (Cheung & Chan, 2012).

Table 6.1 Comparison between truss-out bamboo scaffold and RDP

Criteria	Truss-Out bamboo scaffold	RDP
Material	Bamboo and wooden plank	Steel and aluminium
Durability	Limited number of times	Lasts for five years or more
Structural design	Depends on norm of the industry	Laboratory tested
Anchor bolt	At least three	No
Wall damage	Yes	No
Installer location	Install from outside the building	Install from inside the building
Risk of falling object	High, many loose parts	Low, no loose parts
Training	Experienced bamboo scaffolders	Trained worker
Cost	HKD 2,000–3,000 each (approx. USD 256–385 each)	HKD 6,000 – 10,000 per set (approx. USD 769 – 1,282 per set)
Time	Three hours' installation Two hours' dismantling Two experienced bamboo scaffolders	Installation and dismantle Two hours, respectively One trained worker
Risk involved	High	Low

Source: Cheung & Chan (2012).

Other engineering solutions

One of the safety hazards of RMAA works is that it is difficult to find proper anchorage points for personal protection equipment (PPE), such as safety harnesses, of RMAA workers. Many of the RMAA workers have to compromise their safety by fixing their safety lanyard to the window or bamboo truss-out scaffold, which are not reliable anchorage points. Several engineering devices are in place to solve this problem.

Temporary transportable anchor device

To improve safety of RMAA works, a temporary transportable anchor device can be adopted to provide a temporary anchorage point for PPE of RMAA workers. They can be temporary props to windows or temporary anchorage devices to door frames (Figure 6.19). As shown in Figure 6.19,

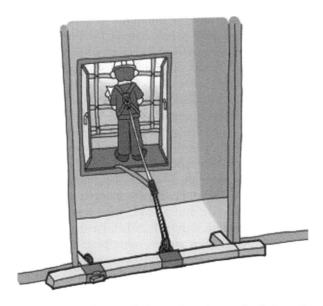

Figure 6.19 Temporary anchorage device to door frames for fixing safety harness.
Source: Occupational Safety and Health Council – Hong Kong (2005).

a temporary anchorage point device is installed at the door frame to provide
an anchorage point for the RMAA worker working at the bamboo truss-out
scaffold to tie the safety harness lanyard.

Podium access system

Besides bamboo truss-out scaffold, many fall from height accidents happened at the ladder. The ladder is often misused by workers to move around
to save time instead of taking the efforts of getting down the ladder, move
the ladder to a desirable location, and climb up the ladder again. To avoid
misuse, some contracting companies do not allow the use of A-frame ladders. The podium access system is an alternative device to replace an A-frame
ladder. A podium access system is safer because its wheels can be fixed and
there are guardrails to protect the worker using it.

Prevention through design

Engineering solutions to improve the safety of RMAA works are temporary
remedies for unavoidable risks. The ultimate goal should be eliminating hazards during the design stage of construction (Gambatese, Behm, & Hinze,
2005). Design for safety or prevention through design is more effective and
economical to improve RMAA safety (NIOSH, 2010). Prevention through
design is a national initiative for the US government to improve construction

safety (NIOSH, 2010). According to NIOSH (2014), 'This approach would incorporate safety features into the building's design, address fall hazards in construction plans, establish safety criteria for buying equipment, and communicate risks to building owners and facilities personnel (Behm, 2005) rather than rely on other forms of protection such as personal protective equipment (PPE) or administrative controls.'

Risks involved in RMAA works can be 'designed out' or minimized should the designers have taken repair and maintenance safety into consideration in their designs. The concept of life-cycle design for safety includes safety during construction, operation, repair and maintenance and demolition. Many accidents could have been prevented if there had been a design change (Cameron, Gillan, & Duff, 2007; Cooke, Lingard, & Blismas, 2008; Hare, Cameron, & Duff, 2006). Permanent features of prevention through design are more cost-effective and reliable than temporarily installed fall from height protection or PPE (NIOSH, 2014). Passive fall from height protection such as a parapet wall is always more preferred than an active fall from height protection such as a fall arrest system or PPE.

Parapet wall

A parapet wall on the rooftop is a recommended building feature to prevent fall hazards (NIOSH, 2013). As shown in Figure 6.20, parapet walls are built on the roof edges and should be at least 39 inches (99 cm) high. Existence of parapet walls helps to prevent workers from fall hazards when they are working too close to the roof edges, such as when they are transferring materials from the roof, accessing rooftop equipment, and communicating with workers on the ground. With built-in parapet walls, temporary fall from height preventive measures can be minimized.

Guardrail embeds

Guardrail embeds are another recommended building feature to prevent fall hazards (NIOSH, 2014). According to NIOSH (2014), edges of concrete slabs could be installed with steel embeds. These steel embeds would facilitate

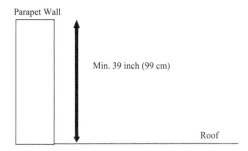

Figure 6.20 Design of a building using a parapet.
Source: Modified from NIOSH (2013).

easy installation of guardrails along the edges of all floors to provide fall protection to workers during the construction process. Instead of using not-so-secured and time-consuming bolted guardrail systems, guardrail installations with the support of the steel embeds are safer and quicker. Embeds are not only useful in providing temporary support for guardrail systems during construction but also useful for permanent guardrail or parapet systems for operation, repair and maintenance.

Roof anchor points

Permanent roof anchor points are recommended to be installed on rooftops (NIOSH, 2014). Anchor points would be useful for fixing a personal fall restraint or fall arrest system. Built-in anchor points can be placed during construction to provide fasten points for safety lanyards and independent lifelines. Permanent roof anchor points should be independent of anchor points for suspended platforms and should be strong enough to support at least 5,000 pounds (2,268 kg).

Chapter summary

To sum up, this chapter has discussed several engineering solutions to improve the safety of RMAA works. RDP is an efficient alternative to truss-out bamboo scaffold for repair and maintenance works. RDP removes the safety hazards involved in putting up and taking off a bamboo truss-out scaffold. There are also other engineering measures currently adopted in the industry such as the podium access system, which replaces the A-frame ladder. Although engineering devices to reduce risks are important, the ultimate solution for RMAA safety should be prevention through design. It is more efficient and economical to embed access for RMAA works, permanent anchorage points, fall arrest systems, etc. early in the design phrase of the construction rather than relying on temporary fall protection or PPE.

References

Behm, M. (2005). Linking construction fatalities to the design for construction safety concept. *Safety Science, 43*, 589–611.

Cameron, I., Gillan, G., & Duff, A. R. (2007). Issues in the selection of fall prevention and arrest equipment. *Engineering, Construction and Architectural Management, 14*(4), 363–374.

Cheung, E., & Chan, A. P. C. (2012). Rapid demountable platform (RDP) – a device for preventing fall from height accidents. *Accident Analysis and Prevention, 48*, 235–245.

Construction Industry Institute – Hong Kong. (2009). *Developing a Prototype for the Rapid Demountable Platform (RDP): Stage II of CII-HK Research on 'Construction Safety Involving Working at Height for Residential Building Repair and Maintenance': Research Summary.*

Cooke, T., Lingard, H., & Blismas, N. (2008). ToolSHeD™: The development and evaluation of a decision support tool for health and safety in construction design. *Engineering, Construction and Architectural Management, 15*(4), 336–351.

Gambatese, J. A., Behm, M., & Hinze, J. W. (2005). Viability of designing for construction worker safety. *Journal of Construction Engineering and Management* (September), 1029–1036.

Hare, B., Cameron, I., & Duff, A. R. (2006). Exploring the integration of health and safety with pre-construction planning. *Engineering, Construction and Architectural Management, 13*(5), 438–450.

National Institute for Occupational Safety and Health. (2010). *Prevention through Design: Plan for the National Initiative: Publication no. 2011-121.*

National Institute for Occupational Safety and Health. (2013). *Preventing Falls through the Design of Roof Parapets: Publication No. 2014-108.*

National Institute for Occupational Safety and Health. (2014). *Preventing Falls from Heights through the Design of Embedded Safety Features: Publication No. 2014-124.*

Occupational Safety and Health Council – Hong Kong. (2005). OSH for Building Services Work. Retrieved from www.oshc.org.hk/others/bookshelf/CB098C.pdf.

7 Measuring safety climate of RMAA works

Introduction

This chapter[1] begins with a discussion of concepts relating to safety climate. Then it discusses the key safety climate factors. It also introduces a new tool for measuring safety climate of RMAA works developed by Hon (2012). Finally, the relationships between safety climate and safety performance of RMAA works (Hon, 2012) are discussed.

Concepts relating to safety climate

Organizational climate and organizational culture

Safety climate is part of organizational climate. A better understanding of organizational climate would help explain the meaning of safety climate. Organizational climate is the 'perception of formal and informal organizational policies, practices, and procedures' (Reichers & Schneider, 1990). Similarly, Ashforth (1985) considers organizational climate to be 'a function of the organization's structure, the people inside the organization and their efforts to understand the organization' (Ashforth, 1985). Ostroff, Kinicki, and Tamkins (2003, p. 566) define organizational climate as 'an experientially based description of what people see and report happening to them in an organizational situation. Climate involves employees' perceptions of what the organization is like in terms of practices, policies, procedures, routines, and rewards'.

Climate and culture are often used interchangeably but they have different meanings. Schein (2010) states that climate is the manifestation of culture. Climate is about individual perceptions but culture is more about beliefs, perceptions, and behaviour. Safety climate and culture are subsets of organizational climate and culture, respectively (Coyle, Sleeman, & Adams, 1995). Safety climate is often considered to be the manifestation of safety culture in the behaviour and perception of employees; however, safety culture is regarded as the context within which individual safety attitudes develop and persist and safety behaviour is promoted (Cox & Flin, 1998).

Safety climate

Zohar (1980) may be the first researcher to investigate safety climate, a subset of organizational climate. He defines safety climate as 'a summary of molar perceptions that employees share about their work environments…a frame of reference for guiding appropriate and adaptive task behaviours' (Zohar, 1980, p. 96). It reflects the true perceived priority of safety in an organization (Zohar, 2003). Griffin and Neal (2000) regard safety climate as the employees' perceptions towards policies, procedures and practices relating to safety. Safety climate is a current-state reflection of the underlying safety culture (Mearns & Flin, 1999; Zohar, 2003).

Safety climate and safety culture

Safety climate and safety culture are hard to delineate. Denison (1996) explains their difference:

> Climate refers to a situation and its link to thoughts, feelings and behaviours of organizational members. Thus, it is temporal, subjective and often subject to direct manipulation by people with power and influence. Culture, in contrast, refers to an evolved context (within which a situation may be embedded). Thus, it is rooted in history, collectively held, and sufficiently complex to resist many attempts at direct manipulation. (p. 644)

Climate and culture are alternative depictions of values and assumptions (Denison, 1996). Schein (2010) suggests that there are three layers of cultural representation. The innermost layer is basic assumptions, the middle layer is espoused values, and the outer layer is artefacts. With reference to these layers of cultural representation, climate is the espoused values and artefacts and culture is basic assumptions. Climate is temporal and it changes quickly. In contrast, culture is stable and it takes a longer time to change.

Safety climate studies often have different research methodology from those of safety culture (DeJoy, Schaffer, Wilson, Vandenberg, & Butts, 2004). Most safety climate research studies would use a questionnaire survey as the research method whereas most safety culture research studies would adopt a qualitative or an ethnographic research methodology (Guldenmund, 2000; Mearns & Flin, 1999). Likewise, Denison (1996) and Guldenmund (2007) consider safety climate and safety culture to be similar concepts but different manifestations for assessing the importance of safety within an organization.

Safety climate, safety attitude and safety perception

Safety climate, safety attitude and safety perception are related concepts sharing some similarities. The term 'safety attitudes' often appears in safety climate research. Safety attitude is often implicitly assumed as safety climate (e.g., Siu, Phillips, & Leung, 2004) or being included as one of the safety climate factors (e.g., Zhou, Fang, & Mohamed, 2011). Guldenmund (2007, p. 726) claims that 'safety climate research is basically attitude research'. Zohar (1980) defines that 'safety climate is the shared safety perceptions of employees towards the organization with which they work'.

Attitude is 'a predisposition to respond in a favourable or unfavourable way to objects or persons in one's environment' (Steers, 1981) and 'a learned tendency to act in a consistent way to a particular object or situation' (Fishbein & Ajzen, 1975, p. 6). These aforementioned definitions suggest that attitudes are learned and nurtured through social interactions and experiences. Attitudes represent the likelihood to act in some ways; however, attitudes cannot perfectly predict actual behaviour. It is believed that attitudes are reasonably consistent and cannot be changed easily. Perception is the process by which people interpret and organize sensation to produce a meaningful experience of the world (Lindsay & Norman, 1972). Thus, perceptions are more susceptible to change when the circumstances have been changed.

Attitudes are personal 'evaluations' of the same object whereas perceptions can be regarded as 'descriptive' and referring to 'external objects' (Guldenmund, 2007). In theory, attitudes and perceptions are separate entities. However, in reality, perceptions are inseparable from attitudes. Thus, Guldenmund (2007) states that 'safety climate is actually attitude research'.

Key safety climate factors in construction

Safety climate is a multidimensional construct. The number of dimensions to form the measurement scale of safety climate varies, often ranging from two to eight. Zohar (1980) initially identifies eight dimensions of safety climate, namely: perceived importance of safety training programs, perceived management attitudes towards safety, perceived effects of safe conduct on promotion, perceived level of risk at the workplace, perceived effects of workplace on safety, perceived status of the safety officer, perceived effects of safe conduct on social status, and perceived status of the safety committee. Similar studies have been conducted subsequently (Brown & Holmes, 1986; Coyle et al., 1995; Dedobbeleer & Béland, 1991), hoping to clearly identify the dimensions of safety climate. However, results are not replicable.

The factor structure of safety climate seems to not be replicable (Coyle et al., 1995) and is industry-specific (Cox & Flin, 1998). Items developed in one industry are not likely to be generalized to other industries. As suggested by Lin, Tang, Miao, Wang, and Wang (2008), these can be due to different populations in different industries or cultures, and different selections of factor analysis procedures. Shannon and Norman (2009) note that inappropriate application of factor analysis is one of the reasons leading to variations of safety climate scales. Data input for factor analysis are usually individual workers' perception on safety management system, practices, etc. The object of measurement in safety climate scale items are, however, the work group or the company, making the scale items multilevel. Individual workers' perceptions are often added together without considering the within-group homogeneity. In that case, multilevel statistical analysis, such as multilevel confirmatory factor analysis, should be more appropriately used to determine the factor structure of safety climate.

Single-level measurement of safety climate has evolved to multilevel. Zohar (2000) has proffered a group-level model of safety climate. Zohar's study is an echo of Hofmann and Stetzer's study (1996), which adopted a cross-level approach of safety climate investigation. Group level safety climate takes into account the variations of safety climate levels between different groups. Since safety practices and policies are often carried out at work-group level by different supervisors, some work groups may have higher levels and strength of safety climate while some do not, even within the same organization.

A number of studies investigating safety climate factors in the construction industry are listed in Table 7.1 (Choudhry, Fang, & Lingard, 2009; Dedobbeleer & Béland, 1991; Fang, Chen, & Wong, 2006; Glendon & Litherland, 2001; Mohamed, 2002; Zhou et al., 2011). The two earliest studies listed in Table 7.1 were undertaken by psychology researchers. The studies of Dedobbeleer and Béland (1991) and Glendon and Litherland (2001) were conducted in the construction industry by using safety climate questionnaires originally developed for other industries. The remaining three studies were conducted by researchers in the construction industry. Mohamed (2002) may be one of the earliest researchers in construction to measure construction safety climate. Fang et al. (2006), Choudhry et al. (2009), and Zhou et al. (2011) are closely related studies contributing to the recent development of safety climate research in the construction industry.

Based on 71 questions of the safety climate questionnaire developed by Health and Safety Executive (2001) and 16 items covering 14 safety management elements complied by the Hong Kong government, Fang et al. (2006) empirically tested the 87-item questionnaire with data from construction sites in Hong Kong, yielding ten key factors. The follow-up study of Choudhry et al. (2009) reduced the number of items in the questionnaire to 22 and the number of factors to two. Zhou et al. (2011) also conducted a closely related study in China in 2004 and 2007 using a shortened version of the questionnaire of Fang et al. (2006), thereby deriving a four-factor structure of construction safety climate and further reducing the questionnaire to 24 items.

As shown in Table 7.1, the three most common construction safety climate factors are management commitment to safety, safety rules and procedures, and workers' involvement in safety. These three factors are believed to be key safety climate factors in the construction industry because they appear in studies involving construction projects of different sizes and nature conducted in different times and places.

Table 7.1 Comparing safety climate factors in the construction industry

Dedobbeleer and Béland (1991)	Glendon and Litherland (2001)	Mohamed (2002)	Fang et al. (2006)	Choudhry et al. (2009)	Zhou et al. (2011)
Management's commitment to safety		Commitment	Safety attitude and management commitment	Management commitment and employee involvement	Management commitment
Workers' involvement in safety		Workers' involvement	Worker's involvement		
	Safety rules Adequacy of procedures	Safety rules and procedures	Improper safety procedures	Inappropriate safety procedure and work practices	Safety regulations
	Personal protective equipment		Safety resources		
	Communication and support	Communication			
		Personal appreciation of risk Appreciation of hazards	Appraisal of safety procedure and work risk		Safety attitude
	Work pressure	Work pressure			

(continued)

Table 7.1 Comparing safety climate factors in the construction industry (*continued*)

Dedobbeleer and Béland (1991)	Glendon and Litherland (2001)	Mohamed (2002)	Fang et al. (2006)	Choudhry et al. (2009)	Zhou et al. (2011)
	Relationships		Supervisor's role and workmates' role		
			Workmates' influence		Safety training and workmates' support
			Safety consultation and safety training		
		Competence	Competence		
		Risk-taking behaviour	Risk-taking behaviour		

Source: Hon (2012).

A new tool for measuring safety climate of RMAA works

Although a few safety climate measurement scales have been developed for some industries, there is no safety climate measurement tool for RMAA works. Provided that safety climate structure is rather industry-specific (Cox & Flin, 1998), a new tool for measuring the safety climate of RMAA works is needed. By doing so, the safety climate factor structure of the RMAA works can be better revealed.

A study on safety climate factors of RMAA works in Hong Kong was conducted by Hon (2012). Hon (2012) initially adopted the Safety Climate Index (SCI) developed by the Occupational Safety and Health Council – Hong Kong to measure the safety climate of RMAA projects in Hong Kong. Respondents were required to rate 38 safety climate statements with a 5-point Likert scale (1 = strongly disagree, 5 = strongly agree). Some private property management companies, maintenance sections of quasi-government developers and their subcontractors, RMAA project teams of general contractors, small RMAA contractors, building services contractors, and trade unions in Hong Kong participated in the questionnaire survey. A total of 662 completed questionnaires which were valid for analysis, including 397 (60.0%) frontline workers, 131 (19.8%) supervisors, 129 (19.5%) managers and 5 (0.6%) not having disclosed their job positions.

As shown in Table 7.2, a concise measurement scale for the safety climate of RMAA works was derived. Three factors encapsulating 22 statements were identified: F1, management commitment to occupational health and safety (OHS) and employee involvement; F2, applicability of safety rules and work practices; and F3, responsibility for health and safety.

Table 7.2 Safety climate factors of RMAA works

Factor 1 (F1) – Management commitment to OHS and employee involvement

- The company really cares about the health and safety of the people who work here.
- There are good communications here between management and workers about health and safety issues.
- The company encourages suggestions on how to improve health and safety.
- I am clear about what my responsibilities are for health and safety.
- I think management here does enough to follow up recommendations from safety inspection and accident investigation reports.
- All the people who work in my team are fully committed to health and safety.
- There is good preparedness for emergency here.
- Accidents which happened here are always reported.
- Most of the job-specific safety trainings I received are effective.
- I fully understand the health and safety risks associated.
- Safety inspection here is helpful to improve the health and safety of workers.
- Staff are praised for working safely.

(continued)

Table 7.2 Safety climate factors of RMAA works (*continued*)

Factor 2 (F2) – *Applicability of safety rules and work practices*

- Some jobs here are difficult to do safely.
- Not all the health and safety rules or procedures are strictly followed here.
- Some of the workforces pay little attention to health and safety.
- Some health and safety rules or procedures are difficult to follow.
- Supervisors sometimes turn a blind eye to people who are not observing the health and safety procedures.
- Sometimes it is necessary to take risks to get the job done.

Factor 3 (F3) – *Responsibility for health and safety*

- People are just unlucky when they suffer from an accident.
- Accident investigations are mainly used to identify who should be blamed.
- Work health and safety is not my concern.
- Little is done to prevent accidents until someone gets injured.

Source: Hon (2012).

F1 and F2 have been most commonly identified in the literature (Choudhry et al., 2009; Dedobbeleer & Béland, 1991; Fang et al., 2006; Mohamed, 2002; Zhou et al., 2011). Management commitment to OHS and employee involvement is an important factor because effective OHS management needs promotion from the top and support from the bottom. Applicable safety rules and work practices prevent potential hazards from endangering the RMAA workers.

For F3, responsibility for health and safety, it has been identified in safety climate studies conducted in industrial gas production (Cox & Cox, 1991), manufacturing (Cheyne, Cox, Oliver, & Tomas, 1998), and offshore oil and gas production (Mearns, Flin, Fleming, & Gordon, 1998). As revealed in Clarke (2000), this is one of the predominant safety climate factors. Hon (2012) has concisely revealed the factor of responsibility for safety and health, which may have been hidden or ubiquitously scattered across several factors in previous construction safety climate studies.

The three-factor structure of the RMAA safety climate revealed in Hon (2012) shares commonalities with other safety climate studies in construction and other industries. Despite this, managing these factors in the context of RMAA works would involve peculiar challenges pertinent to the nature of RMAA works.

As for management commitment to OHS and employee involvement, small- and medium-sized RMAA contracting companies may have inadequate awareness and resources for safety (Hon, Chan, & Wong, 2010). Very often, RMAA worksites are ubiquitously scattered in location. It is particularly difficult for the management to demonstrate its commitment to OHS and enlist employee involvement (Hon, Chan, & Chan, 2011).

For applicability of safety rules and practices, some safety rules applicable to new construction works may not be equally applicable to RMAA works because of their small scale and short duration. For example, the law of Hong Kong requires construction projects with over 100 workers to employ a safety officer; however, this requirement usually does not apply to RMAA projects because they seldom employ more than 100 workers on site (Hon et al., 2011). Although most construction companies have generic method statements for general building works, they cannot directly address the potential risks and problems in RMAA works (Hon et al., 2011). For example, the risk involved in RMAA works undertaken at the external wall of an old building is different from that of a new building because concrete strength of their respective external walls are likely to be different. The challenge would be designing a set of safety rules and good practices to suit the needs of RMAA works.

As for the responsibility for health and safety, RMAA workers working in occupied buildings and handling minute tasks may easily underestimate the importance of safety. RMAA projects undertaken by SMEs may not have a comprehensive safety system, or may be lacking in terms of transparency and communication. Workers can easily develop a mindset that takes safety for granted, or care only when an accident actually occurs. This can lead to a subsequent negative impression on accident investigation.

How does safety climate influence safety performance of RMAA works?

Safety performance

Safety climate is a useful construct because it is theoretically supported to explain, predict and influence safety behaviour and thus safety performance. Before looking at how safety climate would help to improve safety performance, safety performance has to be well defined for measurement.

Selecting reliable and valid safety performance measurement indicators is important. Objective performance indictors appear to be better than subjective ones. However, Muckler and Seven (1992) argued that objectiveness or subjectiveness is not a good criteria for selecting performance measures because all measurement would involve some subjective elements in selecting measures or in collecting, analysing, or interpreting data. Good performance measures should be relatively simple, adequately valid, sufficiently reliable and appropriately precise.

Safety performance measurement techniques can be broadly categorized into statistics, behavioural measures, periodic safety audits and a balanced score card approach (Chan et al., 2005). The latter three take a relatively longer time to prepare and may not be easily measured in the questionnaire (Choudhry et al., 2009).

Previous safety studies tend to use statistical data of accidents or injuries to measure safety performance; however, accidents or injuries are reactive and relatively infrequent. They may not be effective indicators of safety because they reflect only occurrences of failures (Cooper & Phillips, 2004). They are also 'insufficiently sensitive, of dubious accuracy, retrospective, and ignore risk exposure' (Glendon & Litherland, 2001, p. 161).

Lingard, Cooke, and Blismas (2011) found that lost time and medical treatment injury rates occur infrequently and are ineffective indicators of safety performance. They suggest using a more fine-grained measure of work group safety performance, such as micro-accidents or minor (non-reportable) injuries in future research.

Including minor injuries will enlarge the pool of injury data available for analysis because minor injuries often occur more frequently than serious ones (Beus, Payne, Bergman, & Arthur, 2010). According to Beus, Payne, et al. (2010, p. 717), 'safety climate should be more effective in predicting injuries of a less serious nature'.

Some researchers consider safety behaviour a more preferred measure of safety performance. Burke, Sarpy, Tesluk, and Smith-Crowe (2002, p. 432) defined safety performance as 'evaluative actions or behaviours that individuals exhibit in almost all jobs to promote the health and safety of workers, clients, the public, and the environment'. Burke et al.'s (2002) definition of safety performance provides the justification of measuring safety performance by safety behaviour. With reference to Neal and Griffin (2004), safety compliance and safety participations are reliable dimensions of safety performance measurement.

Griffin and Neal (2000) defined safety compliance as 'following rules in core safety activities'. This includes 'obeying safety regulations, following correct procedures, and using appropriate equipment' (Neal & Griffin, 2004, p. 16). It refers to 'the core activities that individuals need to carry out to maintain workplace safety. These procedures include adhering to standard work procedures and wearing personal protective equipment' (Neal & Griffin, 2006, p. 947).

Safety participation refers to 'behaviours that do not directly contribute to an individual's personal safety but that do help to develop an environment that supports safety' (Neal & Griffin, 2006, p. 947). These behaviours include activities such as 'participating in voluntary safety activities, helping coworkers with safety-related issues, and attending safety meetings' (Neal & Griffin, 2006, p. 947).

Table 7.3 shows 37 research articles reviewed on safety performance measurement in relation to safety climate. Table 7.3 was initially built on Clarke and Cooper (2004) and was updated with a systematic search of journal articles in August 2011. A journal article search was done in Google Scholar and Scopus. Articles containing keywords of both 'safety climate' and 'safety performance' were selected. Only those articles which clearly discussed the measurement of safety performance were included. Most of the studies adopted accidents or injuries as safety performance indicators. Cooper and Phillips (2004) pointed out that proactive measures of safety assessing the current safety activities in the workplace would be better indicators. Safety compliance to rules and regulations and the degree of safety participation are possible safety performance measurement dimensions for a questionnaire survey.

Theoretical foundation

Social exchange theory and expectancy-valence theory lay the theoretical foundation to explain and predict the relationship between safety climate and safety behaviour (Neal & Griffin, 2006).

Social exchange theory postulates that, when an organization cares for its employees' well-being, the employees are likely to develop implicit obligations to perform behaviours beneficial to the organization. Apart from their standard core work duties, they also perform organizational citizenship behaviour, that is, extra-role functions other than core work activities. Hofmann and Morgeson (1999) found that when a company puts emphasis on safety, its employees would reciprocate by complying with safety procedures of the company (Neal & Griffin, 2006).

Table 7.3 Safety performance measurement

		Accidents/ self-reported injuries	Safety compliance	Safety participation
1	Brown and Holmes (1986)	✓		
2	Donald and Canter (1994)	✓		
3	Hofmann and Stetzer (1996)	✓		
4	Cree and Kelloway (1997)	✓		
5	Williamson et al. (1997)	✓		
6	Hayes et al. (1998)	✓	✓	
7	Mearns et al. (1998)	✓		
8	Lee (1998)	✓		
9	Zohar (2000)	✓		
10	Garavan and O'Brien (2001)	✓		
11	Giffin and Neal (2000)		✓	
12	Lee and Harrison (2000)	✓		
13	Neal et al. (2000)		✓	✓
14	Probst and Brubaker (2001)	✓	✓	
15	Barling et al. (2002)	✓		
16	Zohar (2002b)	✓		
17	Eklof and Torner (2002)	✓		✓
18	Gillen et al. (2002)	✓		

(*continued*)

Table 7.3 Safety performance measurement (*continued*)

	Accidents/ self-reported injuries	Safety compliance	Safety participation
19 Oliver et al. (2002)	✓		
20 Goldenhar et al. (2003)		✓	
21 Mearns et al. (2003)	✓		
22 Hofmann et al. (2003)			✓
23 Prussia et al. (2003)		✓	
24 DeJoy et al. (2004)			✓
25 Probst (2004)	✓	✓	
26 Siu et al. (2004)	✓		
27 Zohar and Luria (2004)	✓		
28 Michael et al. (2005)	✓		
29 Morrow and Crum (2004)	✓		
30 Wallace and Chen (2005)	✓	✓	
31 Zacharatos et al. (2005)		✓	✓
32 Zohar and Luria (2005)		✓	
33 Clarke (2006a)	✓		
34 Huang et al. (2006)	✓		
35 Neal and Griffin (2006)	✓	✓	✓
36 Lingard et al. (2011)	✓		
37 Lu and Yang (2011)		✓	✓
Total	28	12	8

Source: Hon (2012).

The expectancy-valence theory asserts that employees will follow safety procedures and rules when they expect that they will be rewarded with valued outcomes. When the company values safety, a high level of safety climate exists. Based on behaviour-outcome expectancies, employees are likely to comply with the workplace safety rules and procedures and participate in safety activities under a high level of safety climate workplace environment because they believe that their safety behaviour will be recognized and rewarded (Neal & Griffin, 2006).

Zohar and Luria (2004) state that safety climate is a social-cognitive construct. Thus, people perceive the organizational safety priority from procedures-as-pattern, rather than as discrete procedures. Safety systems and policies do not naturally generate safety. Employees' safety behaviour would be affected by their perception of the true priority of safety in the company. As an example, even though a company may have imposed overt safety policies and management systems, employees will still project a low priority of safety (i.e., a low level of safety climate) when the management values production more than safety under production pressure. Safety climate influences behaviour through behaviour-outcome expectancies (Zohar & Luria, 2003). A low safety climate implies that employees do not put much value on safety, but put more value on short-term gains, such as finishing the work faster. Employees would likely underestimate the likelihood of possible injury in a workplace with a low level of safety climate.

Hypothetical model

Figure 7.1 shows the hypothetical model of relationships between safety climate and safety performance. Injuries (Inj.), safety participation (SP), and safety compliance (SC) were the selected dimensions of safety performance measurement. Three hypotheses were derived from the model. Hypothesis 1 (H1): RMAA safety climate (RMAASC) is negatively related to injuries (Inj). Hypothesis 2 (H2): RMAASC is positively related to safety participation (SP). Hypothesis 3 (H3): RMAASC is positively related to safety compliance (SC).

Empirical relationships

Safety climate is postulated to have a positive relationship with safety performance. However, empirical studies show that such relationship varies in different industrial contexts and with safety performance measurements. Some studies found positive relationships between safety climate and safety performance (Gillen, Baltz, Gassel, Kirsch, & Vaccaro, 2002; Pousette, Larsson, & Törner, 2008; Siu et al., 2004) whereas some could not find any significant relationship between safety climate and safety performance (Cooper & Phillips, 2004; Glendon & Litherland, 2001).

Among studies with significant safety climate and performance relationships, some of them show relatively weak relationships. For example, Neal, Griffin, and Hart (2000) found that safety climate has significant influence on

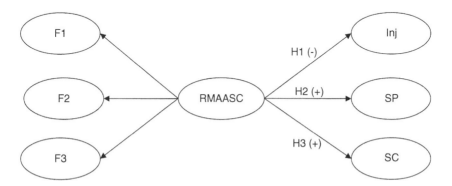

Figure 7.1 Hypothetical model.
Source: Hon (2012).

safety participation (standardized path coefficient = 0.23); however, safety climate does not significantly affect safety compliance. A meta-analysis conducted by Clarke (2006) revealed that safety climate and safety performance are only weakly (0.20) related. Morrow et al. (2010) found a similar result, that safety climate accounts for only 18% of variance in unsafe behaviour.

In contrast, some studies show a relatively strong relationship between safety climate and safety performance. Lu and Yang's (2011) study in the passenger ferry context found that safety climate accounts for 48% and 57% of variance in safety compliance and safety participation, respectively. Two safety climate factors, safety training and emergency preparedness, are significantly related to safety compliance and safety participation. In Larsson, Pousette, and Törner's (2008) study, psychological climate and safety behaviours of construction workers have a significant correlation (r = 0.34). Seo's (2005) study revealed a direct influence of safety climate on safe work behaviour (standardized path coefficient = 0.73) of workers in the US grain industry.

The aforementioned contrasting empirical findings can be explained by several possible reasons. The first reason is the research design and the measurement of safety climate and safety performance. The safety climate and performance relationship may be inflated because of the inclusion of some arguable safety climate factors, such as safety attitudes (Beus, Bergman, & Payne, 2010). Research design may be a moderator of the relationship between safety climate and safety performance (Beus, Bergman, et al., 2010). Clarke (2006) showed that only prospective research design, which measures accidents after measuring safety climate, has validity generalization. In contrast, retrospective design, which measures accident before measuring safety climate, does not have validity generalization. Insensitive safety performance measures, such as objective injury data, have limitations in capturing the variances of safety performance at different levels of safety climate (Lingard, Cooke, & Blismas, 2010). The second reason is that climate strength may moderate the climate–behaviour relationship. With low homogeneity of safety climate perception (i.e., low safety climate strength), the safety climate and safety behaviour relationship becomes weaker (Zohar & Luria, 2004).

The third reason is the level of analysis. Research shows that the relationship between safety climate and safety behaviour varies when their relationships are modelled on different levels of analysis. Findings at the individual, group, or organizational level may be different even with the same data set. For example, Hofmann and Stetzer (1996) have found negative correlations between climate scores and unsafe behaviours at the team level, and, in contrast, positive correlations at the individual level. Recently, safety climate at the group level in the construction industry has drawn attention (Lingard, Cooke, & Blismas, 2009); the importance of safety supervisors and coworkers in enhancing safety performance has been recognized (Lingard et al., 2010, 2011). With all the above possible reasons, the influence of safety climate on safety performance varies across different work settings and environments (Clarke, 2006).

To explore further the influence of safety climate on safety performance, some studies have included antecedents of safety climate into the model of safety climate and safety performance. Several studies have tested the mediating effect of safety climate on organizational factors and safety performance (Barling, Loughlin, & Kelloway, 2002; Neal & Griffin, 2002; Neal et al., 2000; Zohar, 2002a, 2002b). Safety climate has been found to mediate the relationship between organizational climate (Neal et al., 2000) and leadership style (Zohar, 2002a, 2002b) on measures of safety performance.

Techniques of structural equation modelling (SEM) are commonly used, such as confirmatory factor analysis, testing measurement model fit and establishing structural path models. SEM is a confirmatory statistical technique which allows testing of a priori theoretical models (Crowley & Fan, 1997). SEM can determine relationships of a series of interdependent, multiple regression equations at the same time. SEM produces more accurate estimates by considering latent variables and error variance parameters (Byrne, 2009).

Hon (2012) tested the relationships of safety climate with safety performance based on 814 data collected from Hong Kong with SEM techniques. Half of the data was for calibration purposes and half of the data was for validation purposes. Results are shown in Figure 7.2.

Hon (2012) found that the relationship between RMAASC and Inj was significantly negative. RMAASC has a significantly positive relationships with SP and SC respectively. F1, F2 and F3 indirectly affected Inj, SP, and SC. The indirect effect between F2 and SC was the strongest (0.84*0.74 = 0.62 in the calibration sample; 0.89*0.66 = 0.59 in the validation sample) amongst all safety climate factors and safety performance. F2, applicability of safety rules and practices, was the most important factor affecting the level of Inj, SP, and SC.

RMAASC and SC have the strongest relationship (0.74 in the calibration sample; 0.66 in the validation sample) and the strongest explanatory power ($R^2 = 0.55$ in the calibration sample; $R^2 = 0.43$ in the validation sample) as compared with Inj and SP. Thus, SC seems to be the most reliable and valid latent variable of safety performance measurement. It is surprising that the relationship between RMAASC and SP was the weakest, and was in fact even weaker than Inj. This indicates that SP may be affected by variables other than the RMAASC. Perhaps personal safety attitude is one of them.

In the RMAA sector, safety compliance is the safety baseline that needs to be achieved, whereas safety participation is not. Unlike safety compliance, which is considered the obligation of the employee, safety participation involves extra-role activities that are voluntary. Seen in that light, more motivation is needed to perform safety participation than for safety compliance. Perhaps the current level of safety climate in the RMAA sector is strong enough to motivate RMAA workers to comply with safety rules and

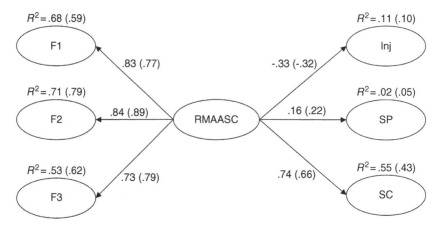

Figure 7.2 Empirically tested structural equation model on the calibration sample
(and the validation sample).
Source: Hon (2012).

regulations; however, it is not strong enough to motivate RMAA workers to participate in extra-role safety activities.

Neal and Griffin (2006) revealed a significant relationship between safety climate and safety participation of hospital employees, but not between safety climate and safety compliance. In contrast, Hon (2012) identified significant relationships between RMAA safety climate and both safety participation and safety compliance. The relationship between RMAA safety climate and safety compliance is particularly strong.

Perhaps such discrepancy can be attributed to the different research design and the industrial settings used. Neal and Griffin (2006) was a longitudinal study in a hospital setting, whereas Hon (2012) was a cross-sectional study in the RMAA sector of the construction industry. It is understandable that a higher standard of safety compliance can be found in a hospital setting, where safety compliance cannot be compromised. Thus, safety compliance is less sensitive to any change in the level of safety climate. Safety participation involving extra-role activities is more sensitive to change in the safety climate. Safety participation is also likely to have a longer time lag for any change in the level of safety climate. Thus, safety participation is better grasped in a longitudinal study. However, the RMAA sector has different circumstances. Safety compliance in RMAA works can be achieved only when a positive safety climate exists. In an RMAA project with a low level of safety climate, the level of safety compliance is likely to be low.

Choudhry et al. (2009) found that the safety climate factor 'management commitment and employee involvement' is more influential on perceived safety performance than another safety climate factor, 'inappropriate safety procedures and work practices'. In contrast, Hon (2012) found that F2, applicability of safety rules and practices, slightly outperforms other factors in having the greatest influence on safety performance, particularly safety compliance, reflecting the peculiar situation of the RMAA sector.

Current safety rules and best practice guidelines of the construction industry are not designed for the RMAA sector. The RMAA sector urgently needs a set of safety rules and practice guidelines that can better meet the specific needs of RMAA works. Proper RMAA safety rules and safety practices should be promulgated and promoted in the RMAA sector.

It is believed that safety performance can be improved by better managing safety climate factors. Problems relating to management procedures and safety systems can be revealed by assessing the level of safety climate (Choudhry et al., 2009). Management commitment to OHS stems from genuine concern for the well-being of the employee. Good safety performance does not occur by chance but is the outcome of proper safety management (Hinze, 2006). Transparent and good communication with workers and supervisors is necessary to convey management's genuine commitment to safety. Safety should be an integral company goal. To enlist employee involvement, workers need to have a clear understanding of their OHS responsibilities and the health and safety risks they will face. They should also be assessed and praised for working safely.

The factor applicability of safety rules and work practices contributes significantly to safety performance. It is dangerous to ask workers to comply with safety rules and work practices that do not fit RMAA works. Safety rules and work practices need to be up-to-date, technically correct, clear (Choudhry et al., 2009) and useful to help the RMAA workers avoid potential risks and hazards, and conduct tasks safely.

To cultivate a higher level of responsibility for health and safety, RMAA workers would need to have a correct evaluation of risk and a locus of control for accidents (Hinze, 2006). Many accidents can be avoided if the workers have ownership for safety. Apart from workers, contracting companies would need to ensure that their employees work in a safe work environment. The purpose of accident investigation is to identify the latent causes of accidents but not find someone to blame. Proactive safety measures are more useful in avoiding injuries or accidents than reactive safety measures.

Chapter summary

This chapter has discussed the key concepts relating to safety climate, a new measurement tool for safety climate of RMAA works, and the effect of safety climate on safety performance of RMAA works. Safety climate is the true priority of safety in a company. It is a useful construct to measure safety of an organization. Safety climate of the RMAA sector in Hong Kong can be measured by a measurement scale of 22 statements. The relationships between safety climate and safety performance of RMAA works were established. Improving safety climate factors of RMAA works would help improve the safety performance of this sector.

Note

1 Extracted from Hon, C. K. H. (2012). Relationships between safety climate and safety performance of repair, maintenance, minor alteration, and addition (RMAA) works. (PhD thesis). Hong Kong Polytechnic University, Hong Kong.

References

Ashforth, B. E. (1985). Climate formation: Issues and extensions. *Academy of Management Review, 10,* 837–847.

Barling, J., Loughlin, C., & Kelloway, E. K. (2002). Development and test of a model linking safety-specific transformational leadership and occupational safety. *Journal of Applied Psychology, 87*(3), 488–496.

Beus, J. M., Bergman, M. E., & Payne, S. C. (2010). The influence of organizational tenure on safety climate strength: A first look. *Accident Analysis and Prevention, 42*(5), 1431–1437.

Beus, J. M., Payne, S. C., Bergman, M. E., & Arthur, W. J. (2010). Safety climate and injuries: An examination of theoretical and empirical relationships. *Journal of Applied Psychology, 95*(4), 713–727.

Brown, R. L., & Holmes, H. (1986). The use of a factor-analytic procedure for assessing the validity of an employee safety climate model. *Accident Analysis and Prevention, 18,* 455–470.

Burke, M. J., Sarpy, S. A., Tesluk, P. E., & Smith-Crowe, K. (2002). General safety performance: A test of a grounded theoretical model. *Personnel Psychology, 55,* 429–457.

Byrne, B. M. (2009). *Structural Equation Modeling with AMOS: Basic Concepts, Applications and Programming* (2nd ed.). Routledge: Taylor and Francis.

Chan, A. P. C., Wong, F. K. W., Yam, M. C. H., Chan, D. W. M., Ng, J. W. S., & Tam, C. M. (2005). *From Attitude to Culture – Effect of Safety Climate on Construction Safety.* Hong Kong, China: The Hong Kong Polytechnic University.

Cheyne, A., Cox, S., Oliver, A., & Tomas, J. M. (1998). Modeling safety climate in the prediction of levels of safety activity. *Work and Stress, 12,* 255–271.

Choudhry, R. M., Fang, D., & Lingard, H. (2009). Measuring safety climate of a construction company. *Journal of Construction Engineering and Management, 135*(9), 890–899.

Clarke, S. (2000). Safety culture: Under-specified and overrated. *International Journal of Management Reviews, 2*(1), 65–90.

Clarke, S. (2006). The relationship between safety climate and safety performance: A meta-analytic review. *Journal of Occupational Health Psychology, 11*(4), 315–327.

Clarke, S., & Cooper, C. L. (2004). *Managing the Risk of Workplace Stress* (pp. 29–60). Great Britain: Routledge.

Cooper, M. D., & Phillips, R. A. (2004). Exploratory analysis of the safety climate and safety behavior relationship. *Journal of Safety Research, 35*(5), 497–512.

Cox, S., & Cox, T. (1991). The structure of employee attitude to safety: A European example. *Work and Stress, 5,* 93–106.

Cox, S., & Flin, R. (1998). Safety culture: Philosopher's stone or man of straw. *Work and Stress, 12*(3), 189–201.

Coyle, I. R., Sleeman, S. D., & Adams, N. (1995). Safety climate. *Journal of Safety Research, 26*(4), 247–254.

Crowley, S. L., & Fan, X. (1997). Structural equation modeling: Basic concepts and applications in personality assessment research. *Journal of Personality Assessment, 68*(3), 508–531.

Dedobbeleer, N., & Béland, F. (1991). A safety climate measure for construction sites. *Journal of Safety Research, 22*(9), 97–103.

DeJoy, D. M., Schaffer, B. S., Wilson, M. G., Vandenberg, R. J., & Butts, M. M. (2004). Creating safer workplaces: Assessing the determinants and role of safety climate. *Journal of Safety Research, 35*(1), 81–90.

Denison, D. R. (1996). What is the difference between organizational culture and organizational climate? A native's point of view on a decade of paradigm wars. *Academy of Management Review, 21*(3), 619–654.

Fang, D., Chen, Y., & Wong, L. (2006). Safety climate in construction industry: A case study in Hong Kong. *Journal of Construction Engineering and Management, 132*(6), 573–584.

Fishbein, M., & Ajzen, I. (1975). *Belief, Attitude, Intention and Behavior: An Introduction to Theory and Research*. Reading, MA: Addison-Wesley.

Gillen, M., Baltz, D., Gassel, M., Kirsch, L., & Vaccaro, D. (2002). Perceived safety climate, job demands, and coworker support among union and non-union injured construction workers. *Journal of Safety Research, 33*, 33–51.

Glendon, A. I., & Litherland, D. K. (2001). Safety climate factors, group differences and safety behavior in road construction. *Safety Science, 39*(3), 157–188.

Griffin, M. A., & Neal, A. (2000). Perceptions of safety at work: A framework for linking safety climate to safety performance, knowledge, and motivation. *Journal of Occupational Health Psychology, 5*(3), 347–358.

Guldenmund, F. W. (2000). The nature of safety culture: A review of theory and research. *Safety Science, 24*, 215–257.

Guldenmund, F. W. (2007). The use of questionnaires in safety culture research – An evaluation. *Safety Science, 45*, 723–743.

Health and Safety Executive. (2001). *Safety Climate Measurement – User Guide and Toolkit*. Offshore Division of the HSE, Chevron UK, Chevron Gulf of Mexico, Mobil North Sea and Oryx UK.

Hinze, J. (2006). *Construction Safety* (2nd ed.). Gainesville, Florida: Alta Systems, Inc.

Hofmann, D. A., & Morgeson, F. P. (1999). Safety-related behavior as a social exchange: The role of perceived organizational support and leader-member exchange. *Journal of Applied Psychology, 84*(2), 286–296.

Hofmann, D. A., & Stetzer, A. (1996). A cross-level investigation of factors influencing unsafe behaviors and accidents. *Personnel Psychology, 49*, 307–339.

Hon, C. K. H. (2012). Relationships between safety climate and safety performance of repair, maintenance, minor alteration, and addition (RMAA) works. (PhD thesis). Hong Kong Polytechnic University, Hong Kong.

Hon, C. K. H., Chan, A. P. C., & Chan, D. W. M. (2011). Strategies for improving safety performance of repair, maintenance, minor alteration and addition (RMAA) works. *Facilities, 29*(13/14), 591–610.

Hon, C. K. H., Chan, A. P. C., & Wong, F. K. W. (2010). An empirical study on causes of accidents of repair, maintenance, minor alteration and addition works in Hong Kong. *Safety Science, 48*(7), 894–901.

Larsson, S., Pousette, A., & Törner, M. (2008). Psychological climate and safety in the construction industry-mediated influence on safety behavior. *Safety Science, 46*(3), 405–412.

Lin, S. H., Tang, W. J., Miao, J. Y., Wang, Z. M., & Wang, P. X. (2008). Safety climate measurement at workplace in China: A validity and reliability assessment. *Safety Science, 46*, 1037–1046.

Lindsay, P., & Norman, D. A. (1972). *Human Information Processing: An Introduction to Psychology*. New York: Academic Press.

Lingard, H., Cooke, T., & Blismas, N. (2009). Group-level safety climate in the Australian construction industry: Within-group homogeneity and between-group differences in road construction and maintenance. *Construction Management and Economics, 27*(4), 419–432.

Lingard, H., Cooke, T., & Blismas, N. (2010). Properties of group safety climate in construction: The development and evaluation of a typology. *Construction Management and Economics, 28*(10), 1099–1112.

Lingard, H., Cooke, T., & Blismas, N. (2011). Co-workers' response to occupational health and safety: An overlooked dimension of group-level safety climate in the construction industry. *Engineering, Construction and Architectural Management, 18*(2), 159–175.

Lu, C. S., & Yang, C. S. (2011). Safety climate and safety behavior in the passenger ferry context. *Accident Analysis and Prevention, 43*, 329–341.

Mearns, K. J., & Flin, R. (1999). Assessing the state of occupational safety – Culture or climate. *Current Psychology, 18*(1), 5–17.

Mearns, K. J., Flin, R., Fleming, M., & Gordon, R. (1998). Measuring safety climate on offshore installations. *Work and Stress, 12*, 238–254.

Mohamed, S. (2002). Safety climate in construction site environments. *Journal of Construction Engineering and Management, 128*(5), 375–384.

Morrow, S. L., McGonagle, A. K., dove-Steinkamp, M. L., Jr., W. C. T., Marmet, M., & Barnes-Farrell, J. L. (2010). Relationships between psychological safety climate facets and safety behavior in the rail industry: A dominance analysis. *Accident Analysis and Prevention, 42*, 1460–1467.

Muckler, F. A., & Seven, S. A. (1992). Selecting performance measures: 'Objective' versus 'subjective' measurement. *Human Factors, 34*(4), 441–455.

Neal, A., & Griffin, M. A. (2002). Safety climate and safety behavior. *Australian Journal of Management, 27*(Special Issue), 67–75.

Neal, A., & Griffin, M. A. (2004). Safety climate and safety at work. In J. Barling & M. R. Frone (Eds.), *The Psychology of Workplace Safety* (pp. 15–34). Washington, DC: American Psychological Association.

Neal, A., & Griffin, M. A. (2006). A study of the lagged relationships among safety climate, safety motivation, safety behavior, and accidents at the individual and group levels. *Journal of Applied Psychology, 91*(4), 946–953.

Neal, A., Griffin, M. A., & Hart, P. M. (2000). The impact of organizational climate on safety climate and individual behavior. *Safety Science, 34*(1-3), 99–109.

Ostroff, C., Kinicki, A. J., & Tamkins, M. M. (2003). Organizational culture and climate. In I. B. Weiner (Ed.), *Handbook of Psychology* (pp. 565–593). Hoboken, NJ: Wiley.

Pousette, A., Larsson, S., & Törner, M. (2008). Safety climate cross-validation, strength and prediction of safety behaviour. *Safety Science, 46*(3), 398–404.

Reichers, A. R., & Schneider, B. (1990). Climate and culture: An evolution of constructs. In B. Schneider (Ed.), *Organizational Climate and Culture* (pp. 21–36). San Francisco: Jossey-Bass.

Schein, E. H. (2010). *Organizational Culture and Leadership* (4th ed.). San Francisco: Jossey-Bass.

Seo, D. S. (2005). An explicative model of unsafe work behavior. *Safety Science, 43*, 187–211.

Shannon, H. S., & Norman, G. R. (2009). Deriving the factor structure of safety climate scales. *Safety Science, 47*(3), 327–329.

Siu, O. L., Phillips, D. R., & Leung, T. W. (2004). Safety climate and safety performance among construction workers in Hong Kong: The role of psychological strains as mediators. *Accident Analysis and Prevention, 36*, 359–366.

Steers, R. M. (1981). *Introduction to Organizational Behavior.* Dallas: Scott Foresman and Company.

Zhou, Q., Fang, D., & Mohamed, S. (2011). Safety climate improvement: Case study in a Chinese construction company. *Journal of Construction Engineering and Management, 137*(1), 86–95.

Zohar, D. (1980). Safety climate in industrial organizations: Theoretical and applied implications. *Journal of Applied Psychology, 65*(1), 96–102.

Zohar, D. (2000). A group-level model of safety climate: Testing the effect of group climate in microaccidents in manufacturing jobs. *Journal of Applied Psychology, 85*, 587–596.

Zohar, D. (2002a). The effects of leadership dimensions, safety climate, and assigned priorities on minor injuries in work groups. *Journal of Organizational Behavior, 23*(1), 75–92.

Zohar, D. (2002b). Modifying supervisory practices to improve subunit safety: A leadership-based intervention model. *Journal of Applied Psychology, 87*(1), 156–163.

Zohar, D. (2003). Safety climate: Conceptual and measurement issues. In J. C. Quick & L. E. Tetrick (Eds.), *Handbook of Occupational Health Psychology* (pp. 123–142). Washington, DC: American Psychological Association.

Zohar, D., & Luria, G. (2003). The use of supervisory practices as leverage to improve safety behavior: A cross-level intervention model. *Journal of Safety Research, 34*(5), 567–577.

Zohar, D., & Luria, G. (2004). Climate as a social-cognitive construction of supervisory safety practices: Scripts as proxy of behavior patterns. *Journal of Applied Psychology, 89*(2), 322–333.

8 Demographic variables and safety of RMAA works

Introduction

This chapter[1] discusses demographic variables which affect safety performance of RMAA works with reference to accident proneness theory. Discussion is based on the questionnaire survey findings of Hon (2012). For details of the questionnaire survey, please refer to Chapter 7.

Accident proneness theory

Accident proneness theory is one of the earliest theories of accident causation (Hinze, 2006). According to McKenna (1983), the concept of accident proneness first appeared in the late 1910s. It postulates that some individuals are more likely to incur an accident than others in the same situation. It assumes that an accident does not occur randomly. Some individuals have permanent characteristics such as personal traits that predispose them to a greater probability of being involved in accidents. However, this concept is controversial and arouses much debate. One of the problems of this concept is that it is unfair to blame individuals for accident proneness but ignore other factors such as deficiencies in health and safety regulations of the workplace (Green, 1991). According to Hinze (2006), accident proneness accounted for only around 10% to 15% of accidents. Visser, Pijl, Stolk, Neeleman, and Rosmalen (2007) conducted an analysis on 79 articles with empirical accident rates to investigate whether accident proneness really exists. They concluded that the phenomenon of accident proneness existed empirically; however, the study was limited by different operationalisations of the concept. Instead of arguing that some people have fixed traits that are more prone to accidents, Dahlback (1991) associated accident proneness with the propensity to take risk. As advocated by Hinze (2006), risk-taking is not a permanent trait; for example, it declines with age. Risk-taking can be influenced through proper motivational techniques. Accident-prone workers may not end up in an accident. Their behaviour can be changed by proper safety training and implementation of safety measures (Hinze, 2006).

Demographic variables and safety climate

Literature review shows that demographic variables play a role in affecting individuals' perceptions of safety and their perceptions of taking risk (Fang, Chen, & Wong, 2006; Goncalves, Silva, Lima, & Meliá, 2008; Gyekye & Salminen, 2009; Siu, Phillips, & Leung, 2003; Zhou, Fang, & Wang, 2008).

Age, marital status, family members to support

Siu et al. (2003) conducted a study to investigate relationships between age, safety attitudes and safety performance of Hong Kong construction workers. Siu et al. (2003) found that there was a curvilinear relationship between occupational injuries and age. That is, the number of injuries would increase with age to some point and then would decrease as age increases. Younger workers tend to have more injuries than older workers. Similarly, Fang et al. (2006) conducted a study to examine the relationships between demographic variables and safety climate. Fang et al. (2006) found that employees who were older, married, and with more family members to support had higher levels of safety climate. This suggested that people with greater social responsibility would be more cautious and tend to avoid risk-taking behaviour.

Education

Education and safety performance have positive relationships. Fang et al. (2006) found that employees' education attainment would affect their level of safety climate. Employees with only primary school education were found to have far less positive perceptions of safety climate than others do. Similarly, Gyekye and Salminen (2009) found that workers with higher education attainment have higher safety perceptions, greater safety compliance and fewer accidents.

Drinking habit

Drinking habit is negatively related to safety climate. Employees who drink alcohol at work had lower safety climate than those who do not (Fang et al., 2006). Alcohol may weaken the judgment of a person; hence, those who drink have a higher chance of getting injured. Drinking at work is not allowed. Those who drink at work simply do not care about their own safety and their workmates' safety. It is common that negative working habits go together with alcoholic consumption at work.

Role of employer

The safety climate levels of employees of subcontractors or joint ventures were found to be lower than that of direct employees (Fang et al., 2006).

This may be because of excessive subcontracting which results in lack of control and low commitment towards the company. Indirect labour has a low commitment to the company and their workmates. The level of safety climate of the casual employees is likely to be lower.

Personal experience

Personal experience would influence one's behaviour. Zhou et al. (2008) formed a Bayesian network–based model to examine the relationship between personal experience and level of safety climate. In Zhou et al. (2008), personal experience had four factors, namely: safety knowledge, education experience, work experience and drinking habits. To influence one's safety behaviour, Zhou et al. (2008) found that the most effective way was to manipulate personal experience factors and safety climate factors at the same time.

Work accident experience

Goncalves et al. (2008) found that work accident experience is positively related to external attribution and unsafe behaviours but negatively associated with internal attributions. The findings of Goncalves et al. (2008) indicated that people with work accident experience would attribute the cause of accident to the external environment but not themselves and are likely to have unsafe behaviours.

Employment tenure

Length of employment tenure is related to safety climate strength (Beus, Bergman, & Payne, 2010). The higher the average tenure, the stronger the safety climate. Companies with high employee turnover rate (i.e., short employment tenure) tend to have weak safety climate strength. It takes time for the new employees to learn about the safety climate of an organization.

Others

Some demographic variables have no significant influence on safety climate. For example, gender, work experience with the company, work experience in the construction industry, injured or not, and smoking habit (Fang et al., 2006). No significant relationships were found between these variables and safety climate (Fang et al., 2006).

Important demographic variables vs. safety climate of RMAA works

Figures 8.1 to 8.12 show the distributions of the respondents' mean safety climate scores by different demographic variables. Results were drawn from Hon (2012). The RMAA safety climate score was calculated by averaging the total score of the 22 variables listed in Table 7.2 constituting RMAA safety climate. Mean safety climate score of the respondents increased with working level. Mean safety climate score of frontline workers was lower than that of the managers (Figure 8.1).

Mean safety climate score of respondents increased with age. Older respondents tended to have a higher mean safety climate score (Figure 8.2).

Male respondents had a higher mean safety climate score than their female counterparts (Figure 8.3). However, it should be noted that only 15 female respondents were in the sample. With such a limited number of female respondents, further investigation is needed before drawing any sensible conclusion.

Married respondents had a higher mean safety climate score than those who are single (Figure 8.4).

In general, respondents with family members to support had higher mean safety climate scores than respondents without any family burden. However, the mean safety climate of respondents with 7 or more family members to support was surprisingly low. Since only 8 respondents had 7 or more family members to support, this finding needs to be interpreted with care (Figure 8.5).

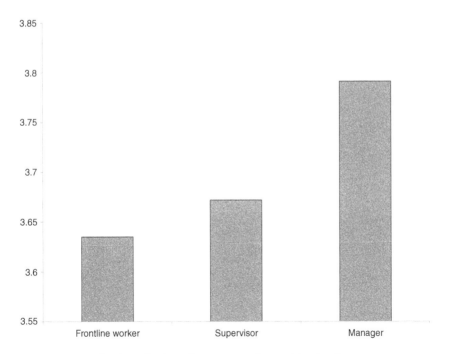

Figure 8.1 Distribution of safety climate scores by working level.

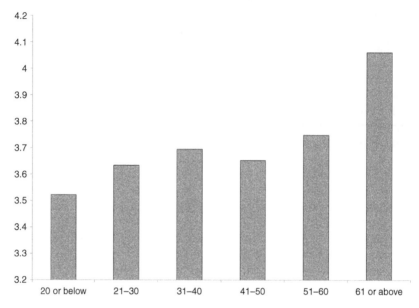

Figure 8.2 Distribution of safety climate scores by age.

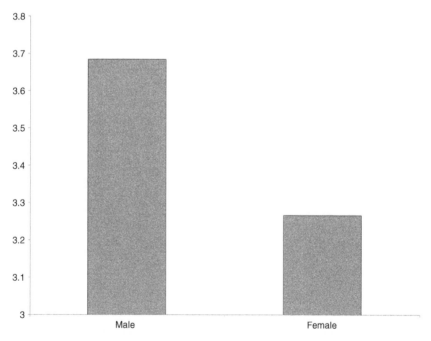

Figure 8.3 Distribution of safety climate scores by gender.

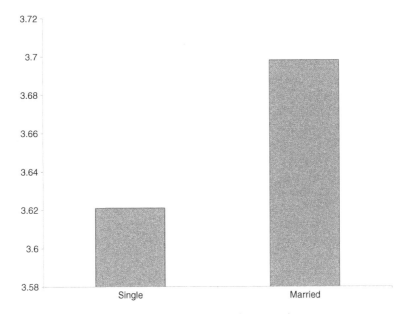

Figure 8.4 Distribution of safety climate scores by marital status.

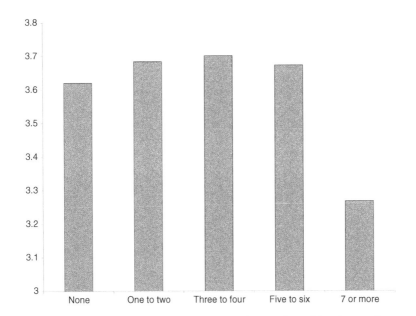

Figure 8.5 Distribution of safety climate scores by number of family members to support.

Respondents with a higher education level had a higher mean safety climate score. The difference is particularly remarkable between primary or below and secondary/certificate/diploma (Figure 8.6).

Respondents directly employed by a client had a higher mean safety climate score than those employed by a main contractor, subcontractor and others. Notably, the mean safety climate score of respondents employed by subcontractors and others was substantially lower (Figure 8.7).

Mean safety climate scores varied with the respondents' length of service in the current company. Newcomers having joined the company for less than a year had the highest mean safety climate. The mean safety climate score dropped when the respondents served the current company for 1 to 5 years (Figure 8.8).

Mean safety climate scores changed with the respondents' years of working experience. The respondents with less than 5 years of working experience had the lowest mean safety climate score whereas those with more than 20 years of working experience had the highest mean safety climate score (Figure 8.9).

As for safety training, it is surprising to find that respondents with a Construction Industry Safety Training Certificate (commonly known as a 'Green Card' in Hong Kong) scored a lower mean safety climate score than those without a Green Card. The respondents with additional safety training qualifications had a higher mean safety climate score (Figure 8.10).

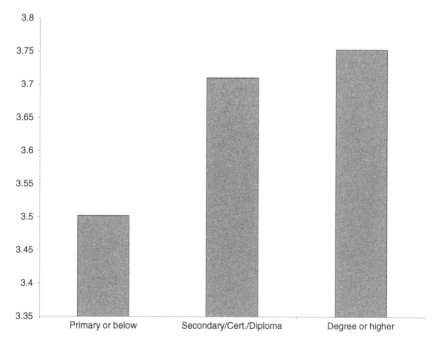

Figure 8.6 Distribution of safety climate scores by education level.

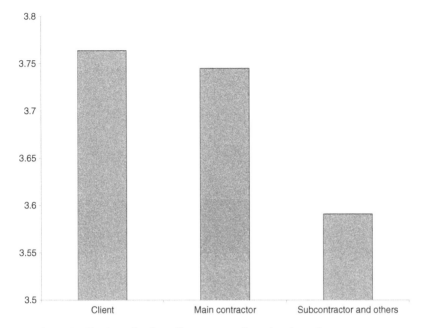

Figure 8.7 Distribution of safety climate scores by role of employer.

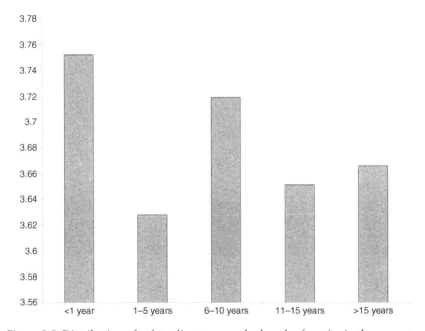

Figure 8.8 Distribution of safety climate scores by length of service in the current company.

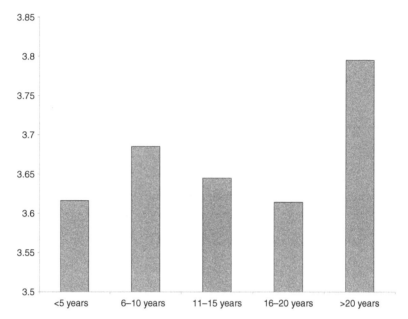

Figure 8.9 Distribution of safety climate scores by working experience.

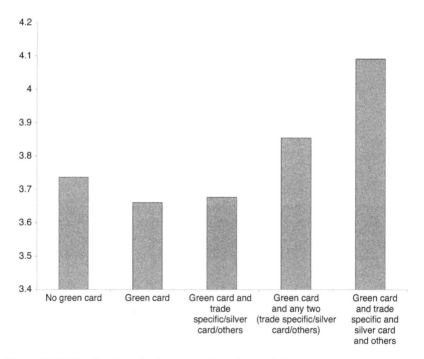

Figure 8.10 Distribution of safety scores by safety training.

The respondents without a smoking or drinking habit had higher mean safety climate scores than those with a smoking or drinking habit (Figure 8.11 and Figure 8.12).

Analysis of Variance (ANOVA) results

Hon (2012) determined whether demographic variables affect the levels of RMAA safety climate by conducting ANOVA analysis between RMAA safety climate scores and demographic variables. As shown in Table 8.1, mean differences of RMAA safety climate scores were significant for respondents with different working levels [$F(2,654)$ = 5.556, p = 0.004], gender [$F(1,655)$ = 10.778, p = 0.001], marital status [$F(1,653)$ = 4.476, p = 0.035], education [$F(2,652)$ = 15.090, p < 0.001], employer [$F(2,643)$ = 10.667, p < 0.001], length of service in the current company [$F(4,649)$ = 2.668, p = 0.031], working experience [$F(4,644)$ = 3.835, p = 0.004], smoking habits [$F(2,654)$ = 4.112, p = 0.017] and drinking habits [$F(2,655)$ = 8.154, p < 0.001].

As shown in Tables 8.1 and 8.2, the frontline workers (M = 3.635, SD = 0.454) and the mangers (M = 3.791; SD = 0.387) had significantly different mean RMAA safety climate scores. This indicates that frontline workers and managers perceived safety differently. Workers' safety perception is not as

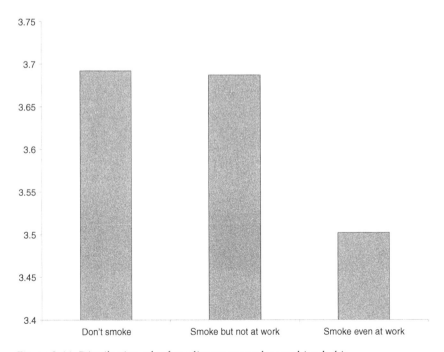

Figure 8.11 Distribution of safety climate scores by smoking habit.

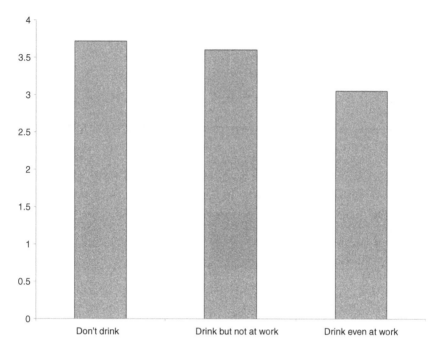

Figure 8.12 Distribution of safety climate scores by drinking habit.

high as that of the managers. Females had significantly lower mean RMAA safety climate scores (M = 3.266, SD = 0.321) than that of the males (M = 3.684, SD = 0.442). However, with only a small sample of female respondents (N = 15), further studies are needed to verify this relationship.

Married people had a significantly higher mean safety climate score (M = 3.698, SD = 0.433) than those who were single (M = 3.621, SD = 0.465). This indicates that people tend to perceive safety to be more important when their sense of responsibility and costs of mishaps increase. People with degrees or higher education had a significantly higher mean RMAA safety climate score (M = 3.753, SD = 0.458) than those who had primary or below level of education (M = 3.502, SD = 0.409). Employees of the subcontractors and others had a significantly lower mean RMAA safety climate score (M = 3.591, SD = 0.403) than that of the employees of the main contractor (M = 3.745, SD = 0.468) and the client (M = 3.764, SD = 0.468). This indicates that safety of the RMAA subcontractors needs more attention.

For those who worked in the current company for less than a year, their mean RMAA safety climate score (M = 3.752, SD = 0.393) was significantly higher than those with 1 to 5 years of experience (M = 3.628, SD = 0.473). This may indicate that safety alertness decreases when people deal with tasks that they are familiar with. A safety training refreshment course should be provided by the RMAA contracting company on a regular basis.

For those who worked in the construction industry for more than 20 years, their mean RMAA safety climate score ($M = 3.795$, $SD = 0.413$) was significantly higher than those of other groups, except the group having 6–10 years of experience.

The mean RMAA safety climate score of those who smoked, even at work ($M = 3.502$, $SD = 0.511$), was significantly lower than those of nonsmokers ($M = 3.692$, $SD = 0.438$). For those who drank, but not at work, their mean RMAA safety climate score ($M = 3.601$, $SD = 0.443$) was significantly lower than those of nondrinkers ($M = 3.716$, $SD = 0.440$). It is not surprising that smoking and drinking habits adversely affect safety. Drinking habits have even stronger negative effect on the safety climate of RMAA workers than smoking habits. The result is in line with Chan, Yam, Chung, and Yi (2012), that drinking habit adversely affect construction workers' physiological responses when they worked under a hot and humid environment. RMAA workers who get drunk the night before going to work are likely to perform their tasks in a less competent and more unsafe manner.

Table 8.1 ANOVA of the demographic variables with the mean RMAA safety climate scores

Demographic variables	Categories	N	M	SD	ANOVA F statistics	Sig.
Working level	Frontline worker	365	3.635	0.454	5.556*	0.004
	Supervisor	126	3.672	0.450		
	Manager	122	3.791	0.387		
Age	20 or below	10	3.522	0.329	1.734	0.125
	21–30	128	3.633	0.471		
	31–40	194	3.694	0.444		
	41–50	198	3.652	0.426		
	51–60	80	3.748	0.449		
	61 or above	3	4.061	0.430		
Gender	Male	598	3.684	0.442	10.778*	0.001
	Female	15	3.266	0.321		
Marital status	Single	194	3.621	0.465	4.476*	0.035
	Married	419	3.698	0.433		

Demographic variables	Categories	N	M	SD	ANOVA F statistics	Sig.
Family members to support	None	81	3.620	0.493	2.103	0.079
	1–2	294	3.683	0.421		
	3–4	199	3.700	0.426		
	5–6	21	3.672	0.553		
	7 or more	8	3.267	0.607		
Education	Primary or below	153	3.502	0.409	15.090*	< 0.001
	Secondary/ cert./diploma	394	3.710	0.437		
	Degree or higher	110	3.753	0.458		
Role of employer	Client	70	3.764	0.468	10.667*	< 0.001
	Main contractor	248	3.745	0.468		
	Subcontractor and others	329	3.591	0.403		
Length of service in the current company	< 1 year	118	3.752	0.393	2.668*	0.031
	1–5 years	249	3.628	0.473		
	6–10 years	98	3.719	0.434		
	11–15 years	71	3.651	0.422		
	>15 years	77	3.666	0.445		
Working experience	< 5 years	125	3.616	0.497	3.835*	0.004
	6–10 years	140	3.685	0.351		
	11–15 years	135	3.645	0.468		
	16–20 years	90	3.614	0.475		
	>20 years	123	3.795	0.413		

(*continued*)

Table 8.1 ANOVA of the demographic variables with the mean RMAA safety climate scores (*continued*)

Demographic variables	Categories	N	M	SD	ANOVA F statistics	Sig.
Safety training	No Green Card	5	3.736	0.494	1.788	0.129
	Green Card	428	3.660	0.442		
	Green Card and trade specific/Silver Card/others	154	3.677	0.433		
	Green Card and any two (trade specific/ Silver Card/ others)	24	3.854	0.500		
	Green Card and trade-specific and Silver Card and others	2	4.091	0.771		
Smoking habit	Does not smoke	380	3.692	0.438	4.112*	0.017
	Smoke but not at work	180	3.687	0.428		
	Smoke even at work	53	3.502	0.511		
Drinking habit	Does not drink	391	3.716	0.440	8.154*	< 0.001
	Drinks but not at work	221	3.601	0.443		
	Drinks even at work	1	3.050	N.A.		

Note: * $p < 0.05$; N = valid number of respondents.; M = mean; SD = standard deviation.

Source: Hon, Chan, & Yam (2012).

Basic Green Card safety training of RMAA workers does not seem to be effective in raising the safety climate of RMAA works. This is in line with the findings of the Construction Industry Institute – Hong Kong (2008),

Table 8.2 Significant results of the ANOVA post hoc tests

Demographic variables	Category (I)	Category (J)	Mean Difference (I–J)	Sig.
Working level	Frontline worker	Supervisor	−0.025	0.839
		Manager	−0.148*	0.003
Education	Primary or below	Secondary/cert./diploma	−0.208*	<0.001
		Degree	−0.250*	<0.001
Role of employer	Subcontractor and others	Client	−0.173*	0.007
		Main contractor	−0.154*	<0.001
Length of service in the current company	< 1 year	1–5 years	0.141*	0.024
		6–10 years	0.037	0.969
		11–15 years	0.114	0.386
		>15 years	0.091	0.605
Working experience	> 20 years	< 5 years	0.179*	0.007
		6–10 years	0.107	0.258
		11–15 years	0.160*	0.023
		16–20 years	0.181*	0.021
Smoking habit	Does not smoke	Smokes but not at work	0.006	0.986
		Smokes even at work	0.173*	0.013
Drinking habit	Does not drink	Drinks but not at work	0.125*	0.001
		Drinks even at work	0.700	0.080

Note: * $p < 0.05$.

Source: Hon, Chan, & Yam (2012).

that the Green Card was the least effective safety initiative; however, the introduction of the Green Card, together with the enactment of safety management regulations have had a significant impact on raising the general safety standard of the industry.

Nine demographic variables significantly affecting the level of RMAA safety climate were found, of which five are in line with Fang et al. (2006). They are as follows: gender, marital status, education level, drinking habits, and direct employer. Fang et al. (2006) stated that the greater the social responsibility, the lower will be the propensity to take risks. Demographic variables would affect safety climate perceptions. Workers who are older, married, and support more family members are likely to have higher mean scores of safety climate (Hon, 2012).

Unmarried RMAA workers who smoke and drink with less than 5 years' work experience in the construction industry and have worked for subcontractors for more than a year are likely to have lower safety climate scores. Subcontractors are usually small in scale, and may not have the resources to provide safety training and safety supervision. This group of people tends to be young and have acquired basic workmanship and skills. When they become familiar with their tasks, they easily lose their alertness towards work safety. They are more prone to accidents; however, their propensity to take risks will change when they become more mature and have more social responsibility to bear. Effective safety measures should be tailored in a way that would change this group's propensity to take risks (Hon, 2012).

Chapter summary

This chapter shows that accident proneness theory plays a role in explaining safety performance of RMAA projects. People with certain demographic variables tend to be more susceptible to accidents. Mean differences of RMAA safety climate were significantly different in nine demographic variables, namely working level, gender, marital status, education, role of employer, length of service in the current company, working experience, smoking habit, and drinking habit. It is hoped that more assistance can be provided to those people who were identified to be more vulnerable in the RMAA sector.

Note

1 Extracted from Hon, C. K. H. (2012). Relationships between safety climate and safety performance of repair, maintenance, minor alteration, and addition (RMAA) works. (PhD thesis). Hong Kong Polytechnic University, Hong Kong.

References

Beus, J. M., Bergman, M. E., & Payne, S. C. (2010). The influence of organizational tenure on safety climate strength: A first look. *Accident Analysis and Prevention, 42*(5), 1431–1437.

Chan, A. P. C., Yam, M. C. H., Chung, J. W. Y., & Yi, W. (2012). Developing a heat stress model for construction workers. *Journal of Facilities Management, 10*(1), 59–74.

Construction Industry Institute – Hong Kong. (2008). *Safety Initiative Effectiveness in Hong Kong: One Size Does Not Fit All*. Retrieved from https://www.irbnet.de/daten/iconda/CIB_DC24446.pdf.

Dahlback, O. (1991). Accident-proneness and risk-taking. *Personality and Individual Differences, 12*(1), 79–85.

Fang, D., Chen, Y., & Wong, L. (2006). Safety climate in construction industry: A case study in Hong Kong. *Journal of Construction Engineering and Management, 132*(6), 573–584.

Goncalves, S., Silva, S., Lima, M. L., & Meliá, J. (2008). The impact of work accidents experience on causal attributions and worker behavior. *Safety Science, 46*, 992–1001.

Green, J. (1991). Accident proneness. *Journal of the Royal Society of Medicine, 84*(8), 510.

Gyekye, S. A., & Salminen, S. (2009). Educational status and organizational safety climate: Does educational attainment influence workers' perceptions of workplace safety? *Safety Science, 47*, 20–28.

Hinze, J. (2006). *Construction Safety* (2nd ed.). Gainesville, Florida: Alta Systems, Inc.

Hon, C. K. H. (2012). Relationships between safety climate and safety performance of repair, maintenance, minor alteration, and addition (RMAA) works. (PhD thesis). Hong Kong Polytechnic University, Hong Kong.

Hon, C. K. H., Chan, A. P. C., & Yam, M. C. H. (2012). Empirical study to investigate the difficulties of implementing safety practices in the repair and maintenance sector in Hong Kong. *Journal of Construction Engineering and Management, 138*(7), 877–884.

McKenna, F. P. (1983). Accident proneness: A conceptual analysis. *Accident Analysis and Prevention, 5*(1), 65–71.

Siu, O. L., Phillips, D. R., & Leung, T. W. (2003). Age differences in safety attitudes and safety performance in Hong Kong construction workers. *Journal of Safety Research, 34*(2), 199–205.

Visser, E., Pijl, Y. J., Stolk, R. P., Neeleman, J., & Rosmalen, J. G. M. (2007). Accident proneness, does it exist? A review and meta-analysis. *Accident Analysis and Prevention, 39*, 556–564.

Zhou, Q., Fang, D., & Wang, X. (2008). A method to identify strategies for the improvement of human safety behavior by considering safety climate and personal experience. *Safety Science, 46*(10), 1406–1419.

9 Strategies for improving safety of RMAA works

Introduction

This chapter discusses various strategies for improving safety of RMAA works identified by Hon, Chan, and Chan (2011) in Hong Kong. These strategies were grounded from interviews with RMAA contractors (Appendix A) and then verified by expert panel members (Appendix B). It is believed that these strategies can be extrapolated to other developed societies where safety of RMAA works has been a rising problem.

Principles for designing safety strategies

Geller (2001) reviewed that traditional paradigm of injury prevention focuses on three 'Es: *Engineer* the safest equipment, environmental settings, and protective devices; *Educate* people regarding the use of the engineering interventions; *Enforce* compliance with recommended safe work practices'. Undoubtedly, these three Es have contributed to improving occupational safety and health. How to achieve continuous safety improvement? Geller (2001) suggests three new Es: *Ergonomics*, *Empowerment*, and *Evaluation* and ten principles of devising company safety strategies:

1   From regulation to corporate responsibility.
2   From failure-oriented to achievement-oriented.
3   From outcome-focused to behaviour-focused.
4   From top-down control to bottom-up involvement.
5   From a piecemeal to a systems approach.
6   From fault-finding to fact-finding.
7   From reactive to proactive.
8   From quick fix to continuous improvement.
9   From priority to value.
10 Enduring values.

These ten principles will help companies design a proactive safety approach with commitment to continuous improvement. Instead of just fulfilling regulatory obligation, company safety strategies formulated with these principles will form part of the corporate social responsibility (Geller, 2001), target for achievement rather than failure avoidance, focus on behaviour rather than injury record and have top to bottom commitment (Geller, 2001).

Strategies for improving safety of RMAA works

According to the findings of Hon et al. (2011), the RMAA sector in Hong Kong mainly implements six types of strategies for improving safety of RMAA works summarised in Table 9.1.

Table 9.1 Strategies for improving safety of RMAA works

Types	Details
1. Incentive and penalty system	• Penalty • Award • Dismissal • Refrain from future tendering • Promotion
2. Legislative control	• Mandatory registration of RMAA contractors and workers • Negligent workers bear legal responsibility
3. Safety management	• Workplace safety planning and risk assessment • Strengthen site safety supervision
4. Safety culture	• Raise safety awareness • Put more safety resources at the beginning of the project • Establish good safety practices right from the start
5. Procurement method	• Select contractors with good safety performance as partners • Safety performance as one of the important criteria for selection of partners • Safety team attends tender interview • Pay for safety scheme
6. Safety training and education	• Relevant safety training • Ganger/leader • Provide safety training to subcontractors

Source: Hon, Chan, & Chan (2011).

Incentives and penalties

Safety incentives and penalties are the most common strategy for improving safety performance in the RMAA sector. Safety incentives can be given as an award or bonus to subcontractors and site teams as recognition of good safety performance. Sometimes safety incentives are in the employment contract as workers' entitlement.

For example, one contractor in Hong Kong includes a safety award/bonus of HKD 1,000 (approximately USD 128) in the employment contract for direct labour. This safety award/bonus will be given provided that the worker is not injured at work in the past 6 months. Safety incentives are also given at site level. Project safety performance of work groups are to be assessed by a site safety officer or project manager every three months. The champion work group will be awarded HKD 1,000 (approximately USD 128) per person. Besides, an addition of HKD 2,000 (approximately USD 256) will be awarded to the champion group for a lunch celebration. Other safety incentives include lucky draw and the best subcontractor award for the year.

An effective award and penalty system has to be carefully designed. According to Hinze (2002), the award system would be more effective if the award is based on good safety behaviour rather than the number of injuries. Incentives should be given out more frequently and to a considerable number of people rather than infrequently and to just a few people. Incentives for the whole team, including both supervisor and workers, tend to be more effective than incentives given only to the supervisor. The award system should be reviewed and revised regularly to sustain interest of the participants. Monetary value of the award should not be too large. Otherwise, workers will be tempted to cover up injury or be non-reporting so as to gain the award for good safety performance. Incentives having been implemented for a long time may be regarded by employees as their normal entitlements. To maintain effectiveness of the safety award system, it has to be constantly evaluated and revised.

Positive safety behaviour and good safety performance should be awarded. By the same token, unsafe behaviour and poor safety performance should be penalized. Penalty can be in the form of an oral warning, written notice and fine. For example, in one medium-sized RMAA contracting company, a penalty of HKD 100 (approximately USD 13) will be imposed on workers who smoke on site and a penalty of HKD 500 (approximately USD 64) will be imposed on workers who do not wear a safety harness. Workers may be fired if they seriously violate safety practices. Unsafe subcontractors will be disqualified for future tendering.

Legislative control

The RMAA sector is often loosely regulated as compared to new construction. Some suggest that RMAA should be a separate trade for licensing of workers under the Mandatory Construction Workers Registration System in

Hong Kong. RMAA workers should be licensed and mandatorily registered. Some suggest that negligent workers should bear part of the legal responsibility. For example, negligent workers may be temporarily suspended from working on site for some time.

One interviewee commented that

> legislation is a passive means; too much legislation may not be good but I think it is still possible [to resort to legislative control]. At present, it is the main contractor that bears most of the legal responsibilities if an accident happens. Only in some cases does the legal liability extend to subcontractor but not at all to the injured worker.

Site safety management

Site safety management plays a crucial role in improving safety of RMAA works. It is likely that RMAA contractors do not have complete control over the RMAA work site environment. Therefore, it is extremely important to have proper workplace safety planning to identify potential hazards and tackle ad hoc problems, and more importantly, devise safety procedures for the workers to follow. Safety management should also carried out pre-worksite briefing and risk assessments of the tasks to be done.

Another important aspect of site safety management is safety supervision. Safety supervision should be strengthened to deter unsafe behaviours and prevent accidents. A practical suggestion was given by one interviewee:

> We provide sufficient safety training to the direct labour and then they could act as a ganger or leader to enforce safety when working together with subcontractors' workers. Safety training is also provided to workers of the subcontractors. Everyone in the team can play the role of a safety supervisor by stopping unsafe behaviours of their counterparts.

Safety culture

Building up good safety culture and raising safety awareness are suggested by many interviewees. Management commitment to improve safety is crucial in creating good company safety culture. In order to show the management's commitment to safety, it is suggested that top management should invest more than required or at least sufficient resources on safety at the initial stage of a RMAA project. It is always easier and more cost-effective to establish a strong positive safety culture right from the start than to rectify negative safety culture later, when safety performance is not up to expectation.

Open communication is part of good safety culture. Companies should set up a platform for workers to communicate safety-related information.

Possible actions include: collecting near-miss accident report cards, issuing a monthly safety bulletin and sharing lesson learnt from incident investigation. The purpose is not to blame unsafe behaviours of workers but to find out the facts and avoid future incident occurrence. Good company safety culture can be cultivated by providing sufficient safety training and developing a company safety campaign for workers to have commitment and a sense of belonging to company. Safety performance should be a recognized criterion for salary review and job promotion.

Procurement method

Procurement has been considered as one of the strategies to improve safety performance. Safety should be one of the selection criteria of awarding contracts to subcontractors. A contractor could select subcontractors with good track records of safety performance and develop partnerships with them. RMAA contractors may select preferred subcontractors which value safety as their first priority.

RMAA contractors may consider putting safety policies into the contract, such as a pay-for-safety scheme to share resources with subcontractors for better safety improvement. RMAA contractors may invite a safety management team (of the client/main contractor) to attend a tender interview to brief the safety standards required for the project. By doing so, tenderers can understand the safety expectations of the project required by the client and main contractor and include necessary safety resources in their revised bids. It is hoped that the chance of cutting corners for safety of RMAA subcontractors can be minimized.

Safety training and education

Specific safety training for RMAA works would be needed. RMAA works involve danger and safety hazards different from those of new construction works. Current construction safety training focuses on new construction works and thus it does not meet the specific needs of RMAA workers. Apart from basic safety training for construction workers, RMAA workers would need more specific safety training to better understand safety hazards involved in typical RMAA works.

Some interviewees advocated that the level of safety training for workers in the RMAA sector should be higher than that in the new construction sector. Other than the Construction Industry Safety Training Certificate (commonly known as the 'Green Card' in Hong Kong), RMAA workers would require additional safety education on how to complete multiple tasks safely. As discussed in Section 5.3.2, safety supervision for RMAA works is more difficult than for new construction works; RMAA workers need to have a higher level of self-requirement and standard of safety. One interviewee stated that (Hon et al., 2011) safety problems of RMAA works are not the same as new works. RMAA workers need specific safety training to perform RMAA works. To cite an example, erecting a bamboo truss-out scaffold with steel brackets from inside of an existing premise is more difficult than to install a massive bamboo scaffold outside for a new building.

Ranked importance of strategies for improving safety of RMAA works

The above-mentioned strategies were ranked by experts mentioned in Appendix B. Relative importance of these strategies is shown in Table 9.2.

Table 9.2 Ranked importance of strategies for improving safety of RMAA works

		Round 1		Round 2	
		Mean	*Rank*	*Mean*	*Rank*
1.	Select RMAA subcontractors with good track record of safety performance.	4.00	5	4.54	1
2.	Raise safety awareness of RMAA workers.	3.69	10	4.54	1
3.	Relevant safety training for specific trades of RMAA works.	3.85	8	4.23	3
4.	Build up good company safety culture.	4.23	4	4.23	3
5.	Safety promotion and education towards RMAA sector.	3.92	6	4.23	3
6.	Strengthen site monitoring and safety supervision.	4.31	2	4.15	6
7.	Clear safe working procedures and guidance for RMAA workers.	3.85	8	4.08	7

		Round 1		Round 2	
		Mean	*Rank*	*Mean*	*Rank*
8.	Provide sufficient safety equipment for RMAA workers (e.g., personal protective equipment (PPE)).	3.62	12	4.08	7
9.	Design for safety of RMAA works.	3.69	10	4.08	9
10.	A mandatory licensing system for RMAA workers.	3.62	12	3.85	10
11.	Award and penalty scheme.	4.54	1	3.77	11
12.	Improvement of site tidiness and housekeeping.	4.31	2	3.77	11
13.	Legislative control.	3.46	14	3.77	13
14.	Implement pay for safety scheme for RMAA works.	3.23	15	3.62	14
15.	Technology innovations for better safety.	3.92	6	3.54	15

With reference to Table 9.2, 'select RMAA subcontractors with good track record of safety performance' and 'raise safety awareness of RMAA workers' are considered the two most important strategies to improve RMAA work safety. The next three important ones are 'relevant safety training for specific trades of RMAA works', 'build up good company safety culture' and 'relevant safety training for specific trades of RMAA works'.

The Hong Kong Construction Industry Review Committee (2001) enlisted selecting subcontractors with good safety performance track record as one strategy to improve safety performance of the construction industry of Hong Kong. Time, cost and quality are the iron triangle of project management. Smallwood and Lingard (2009) suggested that safety should be given the same importance as these three aspects in project management. To increase the status of safety in project management, safety performance should be properly evaluated by reliable indicators (Smallwood & Lingard, 2009).

Selecting RMAA subcontractors with good safety performance needs to be implemented with pertinent procurement arrangement and mechanisms for tender selection. Instead of awarding contract to the lowest-bid tenderer, past safety performance of tenderers should be a key assessment criterion. In fact, safety should be incorporated into supply chain management to achieve socially responsible buying and contracting (Smallwood & Lingard, 2009). Socially responsible construction companies should consider safety performance of their suppliers and subcontractors. For example, large RMAA contractors may select a list of approved suppliers and subcontractors as designated supply-chain partners. Large RMAA contractors may also provide safety advice and assist small- and medium-sized suppliers and subcontractors develop their safety competency (Smallwood & Lingard, 2009).

Many RMAA subcontractors in Hong Kong are small-to-medium-sized companies with mixed levels of safety capability and they are slackly regulated (Hon, Chan, & Wong, 2010). It is particularly vital for the RMAA contractor to select RMAA subcontractors with good track records of safety performance. Some large RMAA contractors would develop partnering relationship with a few subcontractors which have excellent safety performance records.

'Raise safety awareness of RMAA workers' is perceived to be the most important strategy for improving safety of RMAA works. While raising safety awareness of construction workers is known to be crucial, it is even more important for RMAA workers. Short duration and small scale project in dispersed locations hinder close site safety supervision of RMAA works. In this light, the safety of RMAA workers heavily rests on workers' safety awareness and sense of safety ownership. Raising the safety awareness of workers is easier said than done. Safety awareness is the intrinsic attitude of one's mindset towards safety. Evidence shows that it is more difficult to change one's intrinsic attitude than extrinsic behaviour (Geller, 2001). One's mindset and attitude towards safety could be gradually changed over time by education and training (Mahalingam & Levitt, 2007). In contrast, reward and penalty instantly change one's safety behaviours but they would soon lose their effects. Geller (2001) suggested that mixed strategies should be adopted to change the employees' extrinsic behaviours and intrinsic attitude. Extrinsic behaviours and intrinsic attitude are interrelated. Extrinsic behaviour change will give the synergy to change intrinsic attitude and vice versa (Geller, 2001).

Safety management system with empowered culture to the workers' level is essential. The key successful factors for successful safety strategies are leadership, commitment of the management, and empowerment of workers in the process of safety management. Besides raising safety awareness of

RMAA workers, construction companies should also establish good safety culture. Safety should be the core value and social responsibility of any key project stakeholders in the RMAA sector (Smallwood & Lingard, 2009).

The strategy 'technology innovation for better safety' is perceived to be the least important. This reflects the current thinking of the industry practitioners in the RMAA sector. RMAA projects usually rely more on handicraft and workmanship but less on heavy equipment. Technology innovation may not seem to play an important role in the context of RMAA works. However, there are huge potentials of wider application of technology to change the existing practice of RMAA works and improve safety. For example, robotics can be very useful in safety inspection, and mobile computing can track the location of RMAA workers in case accidents happen.

Another strategy ranked to be the second least important is 'implement pay for safety scheme for RMAA works'. Pay for safety scheme (now Pay for Safety and Environment Scheme (PFSES)) is one of the initiatives of the Hong Kong government to improve safety performance of the overall construction industry (Hong Kong Government, 2003). PFSES has been adopted generally in government or quasi-government new capital construction projects. PFSES is a safety policy driven by the Hong Kong government that an agreed percentage of the total contract sum has been reserved as an incentive for the contractor to perform safety. At the moment, the effect of PFSES to the RMAA sector is rather limited because it may be inappropriate or insignificant for RMAA projects with small contract value and short project duration to adopt PFSES.

PFSES now applies to all public capital works contracts (Works Branch of Development Bureau, 2008), engineering and mechanical contracts and design-and-build contracts with contract value of HKD 20M (approximately USD 2.6M) or above, and also term contracts with the contract value of HKD 50M (approximately USD 6.4M) or above. PFSES is excluded from all term contracts exclusively for maintenance works (e.g., some E&M maintenance contracts) and contracts completed within 6 months. Because of small contract value and short project duration, PFSES is excluded from many government RMAA projects. The private sector has yet considered PFSES to be an important strategy for their RMAA projects. Safety incentives, like PFSES, are effective only in the short term. Safety motivation will soon die out as people gradually regard the safety incentive as their usual right (Gangwar & Goodrum, 2005).

The third least important strategy for improving safety performance of RMAA works is 'legislative control'. Tighter legislative control may seem to be a quick fix to unsafe behaviours. The key issue is how to enforce the tightened law. There are also some saying that contractors should not bear the all the responsibilities when accidents happen. Negligent workers should be held responsible as well. This claim cannot stand. Not only it is unfair to blame the worker when there could be underlying causes contributing to the accident but also it is difficult for the Labour Department to trace back and justify that the worker who was injured or died in an accident was negligent.

Concluding remarks

Effective strategies for improving safety of RMAA works should be multi-faceted. Based on Hon et al. (2011), strategies for improving RMAA safety should blend traditional three Es, 'engineer, educate, and enforce' together with three new Es, 'ergonomics, empowerment, and evaluation' suggested by Geller (2001).

The most important strategies for improving the safety of RMAA works are related to the thread of providing an environment to develop a positive safety culture and improving safety competency and awareness of RMAA workers. This can be achieved by setting safety as a key criterion for awarding contracts, selecting partner subcontractors with good safety records, establishing safety cultures and providing safety training to RMAA workers. Eventually, workers would need to develop a sense of ownership and responsibility for safety. At the same time, companies need to have safety supervisors to enforce safety practices and evaluate safety performance regularly.

Further studies

Safety of RMAA works is a global issue. However, existing research is scant. Further studies can be carried out in major developed societies to better reveal safety problems of RMAA works and promulgate pertinent preventive measures. It is expected that RMAA works will continue to increase in volume and so will safety problems.

Safety and health issues of green refurbishment and retrofitting works is one of the research priorities set out in the European Agency for Safety and Health at Work (2013). Safety of green refurbishment and retrofitting works requires more attention of the industry, researcher and the community.

Building is a major source of greenhouse gas emission (Climate Works Australia, 2010). Construction of green buildings with better energy efficiency helps reduce greenhouse gas emission. Construction of new green buildings may account for only a small percentage of all the buildings in developed societies each year. Greenhouse gas emission may also be reduced by a small percentage. In contrast, green refurbishment and retrofitting of existing buildings has a huge potential for reducing greenhouse gas emission (Climate Works Australia, 2010).

In order to meet the target of using more renewable energy and reducing the amount of greenhouse gas emission, it is anticipated that more and more green refurbishment and retrofitting works are to be done for improving building energy efficiency, such as green roofing, home insulation and solar panel installation (Charted Institute of Building, 2011). These green projects are newly emerged ones which may not have existed decades ago.

They may involve safety hazards that the industry practitioners are not familiar with (Hanger, 2014). For example, workers may not know that materials used for insulation may be harmful to their health. Work practices are not standardized and workers may not have received relevant training. Unless safety issues of green refurbishment and retrofitting workers are properly tackled, green refurbishment and retrofitting works could not be widely implemented.

Chapter summary

This chapter has examined various strategies for improving safety of RMAA works. Raising the safety awareness of RMAA workers and selecting RMAA subcontractors with good track records of safety performance are perceived to be the most important strategies for improving the safety of RMAA works in Hong Kong. Other important strategies include relevant safety training for specific trades of RMAA works, building up a good company safety culture, and safety promotion and education towards the RMAA sector. Should these strategies be consolidated and properly implemented in the RMAA sector, they may crush the root of the safety problems. Although these strategies are devised from a research based in Hong Kong, they are likely to be applicable to other jurisdictions as well. With the anticipated expansion of RMAA works, especially green retrofitting, more efforts should be devoted to improving the safety of this sector.

References

Charted Institute of Building. (2011). CIOB Carbon Action 2050: *Buildings under Refurbishment and Retrofit*. Bracknell, UK: The Chartered Institute of Building.

Climate Works Australia. (2010). *Low Carbon Growth Plan for Australia*. Clayton, Victoria, Australia: Climate Works Australia.

European Agency for Safety and Health at Work. (2013). *Priorities for Occupational Safety and Health Research in Europe: 2013–2020*. Luxembourg: Publications Office of the European Union.

Gangwar, M., & Goodrum, P. M. (2005). The effect of time on safety incentive programs in the US construction industry. *Construction Management and Economics*, 23(8), 851–859.

Geller, E. S. (2001). *Working Safe: How to Help People Actively Care for Health and Safety*. USA: CRC Press.

Hanger, I. (2014). *Report of the Royal Commission into the Home Insulation Program*. Australia: Royal Commission into the Home Insulation Program. Retrieved from http://apo.org.au/files/Resource/hirc_reportoftheroyalcommissionintothe homeinsulationprogram_sep_2014.pdf.

Hinze, J. (2002). Safety incentives: Do they reduce injuries? *Practice Periodical on Structural Design and Construction*, 7(2), 81–84.

Hon, C. K. H., Chan, A. P. C., & Chan, D. W. M. (2011). Strategies for improving safety performance of repair, maintenance, minor alteration and addition (RMAA) works. *Facilities*, 29(13/14), 591–610.

Hon, C. K. H., Chan, A. P. C., & Wong, F. K. W. (2010). An empirical study on causes of accidents of repair, maintenance, minor alteration and addition works in Hong Kong. *Safety Science*, 48(7), 894–901.

Hong Kong Construction Industry Review Committee (HKCIRC). (2001). *Construct for Excellence. Report of the Construction Industry Review Committee.* Retrieved from www.devb.gov.hk/filemanager/en/content_735/reporte.pdf.

Hong Kong Government. (2003). *Hong Kong Yearbook 2003 Chapter 12 Land Public Works and Utilities.* Retrieved from www.yearbook.gov.hk/2003/english/chapter12/12_00.html.

Mahalingam, A., & Levitt, R. (2007). Safety issues on global projects. *Journal of Construction Engineering and Management, 133*(7), 506–516.

Smallwood, J., & Lingard, H. (2009). Occupational health and safety and corporate social responsibility. In M. Murray & A. Dainty (Eds.), *Corporate Social Responsibility in the Construction Industry* (pp. 261–286). London: Spon Press.

Works Branch of Development Bureau. (2008). *Environment, Transport and Works Bureau Technical Circular (Works) No. 19/2005.* Retrieved from www.devb.gov.hk/UtilManager/tc/C-2005-19-0-1.pdf.

Appendix A
Details of interviews

Interviews were conducted between December 2008 and February 2009 in Hong Kong. Since RMAA contractors are likely to be in close proximity to accidents, invitations were sent to 17 RMAA contractors which were on the approved contractors' list of a property management company in Hong Kong. Interview requests were also extended to industrial contacts of the research team.

Eight RMAA contracting companies acceded to our research interview request. Face-to-face interviews were conducted with senior management representatives of these companies. Each interview lasted for about an hour. I received one written reply from an interviewee.

Each interview was tape-recorded and the transcript was written for later coding of data. As shown in Table A.1, interviews with RMAA contractors undertaking large-sized (around USD 10 million), medium-sized (around USD 1 million) and small-sized (USD 1,000 to less than USD 1 million) RMAA projects in Hong Kong were conducted.

Table A.1 Background of the interviewees

Position of interviewees	Companies' project scale
Director	Large
Project Safety Manager & Project Manager	Large
Managing Director & Senior Manager	Large
Executive Director	Medium
Managing Director	Medium
General Manager	Medium
Senior Project Manager	Small
Director	Small
Vice President (Project Development)	Small

Source: Hon, Chan, & Chan (2011).

Reference

Hon, C. K. H., Chan, A. P. C., & Chan, D. W. M. (2011). Strategies for improving safety performance of repair, maintenance, minor alteration and addition (RMAA) works. *Facilities*, 29(13/14), 591–610.

Appendix B
Details of Delphi survey

A Delphi survey is defined by Linstone and Turoff (1975) as 'a method for structuring a group communication process so that the process is effective in allowing a group of individuals, as a whole, to deal with a complex problem'. A Delphi survey is a method often used in prioritizing a list of items (Okoli & Pawlowski, 2004). It allows independent thinking of the experts, giving feedback and chances for changing their opinions after taking into account others' opinions. Results are more reliable with group consensus.

A Delphi survey was adopted in Hon, Chan, and Wong (2010); Hon, Chan, and Chan (2011) and Hon, Chan, and Yam (2012) because participation of stakeholders with insights into RMAA work safety practice was needed. General practitioners in the construction industry may not be familiar with specific safety problems of the RMAA sector. A Delphi survey allows group decision-making of experts to achieve a group consensus result which is more reliable than the separate evaluations of individuals.

A profile of the expert panel members is shown in Table B.1. The expert panel consisted of representatives from the regulatory department of the Hong Kong government, quasi-government organization and the private sector. Some experts are also serving on the board of the construction safety committee of the Hong Kong government.

Design of the first round Delphi survey was mainly based on the categories identified from interviews and supplemented by literature (Construction Industry Institute – Hong Kong, 2007). In the first Delphi survey exercise, the expert panel members rated the relative importance of causes of RMAA accidents, difficulties of implementing safety practices in the RMAA sector and strategies for improving safety of RMAA works with a 5-point Likert scale (1= least important, 5 = most important) via an online system. Then group results were presented to the expert panel members in real time. The panel members were then given another opportunity to review and revise their answers in the second Delphi survey exercise. After two rounds of Delphi survey exercises, prioritized lists with group consensus for causes of RMAA accidents (refer to Table 5.1), difficulties of implementing safety practices in the RMAA sector (refer to Table 5.3) and strategies for improving safety of RMAA works (refer to Table 9.2) were derived.

Table B.1 Background of the expert panel members

Position of the expert panel members	Organization
Safety Manager	Contractor
Technical Manager	Property management company
Deputy Chief Occupational Safety Officer	Hong Kong Government
Senior Manager (Safety and Health)	Hong Kong Government
Representative	Self-regulatory body of insurers
Manager	Contractor
General Manager	Quasi-government body
Principle Consultant	Occupational Safety and Health Council
Chairmen	Construction Industry Institute – Hong Kong
Manager	Private developer
Senior Structural Engineer	Hong Kong Government
Executive Director	Electrical & mechanical contractor
Safety, Health, Environment & Quality Manager	Utility service company

Source: Hon, Chan, & Chan (2011).

References

Construction Industry Institute – Hong Kong. (2007). *Construction Safety Involving Working at Height for Residential Building Repair and Maintenance: Research Summary*. Research Report No. 9, ISBN 978-988-99558-1-6.

Hon, C. K. H., Chan, A. P. C., & Chan, D. W. M. (2011). Strategies for improving safety performance of repair, maintenance, minor alteration and addition (RMAA) works. *Facilities*, *29*(13/14), 591–610.

Hon, C. K. H., Chan, A. P. C., & Wong, F. K. W. (2010). An empirical study on causes of accidents of repair, maintenance, minor alteration and addition works in Hong Kong. *Safety Science*, *48*(7), 894–901.

Hon, C. K. H., Chan, A. P. C., & Yam, M. C. H. (2012). Empirical study to investigate the difficulties of implementing safety practices in the repair and maintenance sector in Hong Kong. *Journal of Construction Engineering and Management*, *138*(7), 877–884.

Linstone, H. A., & Turoff, M. (1975). *The Delphi Method: Techniques and Methods*. Reading, MA: Addison-Wesley Publishing Company, Inc.

Okoli, C., & Pawlowski, S. D. (2004). The Delphi method as a research tool: An example, design considerations and applications. *Information and Management*, *42*, 15–29.

Index